Introduction

After its resurrection in 1953 under the new number OKB-51 (***op**ytno-kon**strook**torskoye by**uro*** – experimental design bureau) the Moscow-based design office headed by Pavel O. Sukhoi started work in two main areas, developing a supersonic tactical fighter for the Soviet Air Force (VVS – Vo***yen**no-voz**doosh**nyye **seel**y*) and a supersonic interceptor for the Air Defence Force (PVO – ***Pro**tivovoz**doosh**naya o**boron**a*). At that time two distinct design schools, or scientific methods, existed in the Soviet Union with regard to high-speed flight. One was headed by Professor Vladimir V. Stroominskiy, an avid proponent of swept wings; the other was led by Professor Pyotr P. Krasil'shchikov, who favoured delta wings. Both types of wings had their merits, therefore OKB-51 worked in both directions at once, designing the single-engined S-1 tactical fighter and the T-3 interceptor which featured considerable structural and systems commonality; the S and T stood for *strelo**vid**noye kry**lo*** (swept wings) and *treu**gol'**noye kry**lo*** (delta wings) respectively. The two aircraft eventually evolved into the famous Su-7 *Fitter-A* fighter-bomber and the Su-9 *Fishpot-B* missile-armed interceptor (known in-house as the T-43) respectively, the latter type entering service in October 1960. The Su-9 further evolved into the T-47 interceptor featuring a more powerful fire control radar and longer-range missiles, which entered service in February 1962 as the Su-11 *Fishpot-C*.

In the second half of the 1950s the Western world began fielding new airborne strike weapons systems, forcing the Soviet Union to take countermeasures. In particular, new state-of-the-art interceptors possessing longer range and head-on engagement capability were required for defending the nation's aerial frontiers. Creating such an aircraft appeared a pretty nebulous perspective, considering that many a promising programme for the re-equipment of the VVS and the PVO was terminated when Nikita S. Khrushchov was head of state.

In this generally troubled climate the outlook for the Sukhoi OKB seemed quite favourable at first. however, By mid-1961 it became obvious that the first-line units of the VVS and the PVO had run into big problems with the Su-7B and Su-9, the appallingly low reliability of the Lyul'ka AL-7F afterburning turbojet being one of the worst. In the first 18 months of service, more than 20 aircraft were lost in accidents, more than half of which were caused by engine failures. The PVO began lobbying for the production entry of the Yakovlev Yak-28P *Firebar* interceptor on the grounds that it was twin-engined, ergo safer. The State Committee for Aviation Hardware (GKAT – *Gosu**dars**tvennyy komi**tet** po aviatsi**on**noy tekh**nik**e*) amended its production plans accordingly, and in the three years to follow GKAT's aircraft factories were to manufacture only twin-engined interceptor types – the Yak-28P and the Tupolev Tu-128 *Fiddler* long-range heavy interceptor. Additionally, the rival OKB-155 headed by Artyom I. Mikoyan had begun trials of the promising MiG-21PF *Fishbed-D* light tactical fighter/interceptor powered by a single Tumanskiy R11F2-300 afterburning turbojet which, though less powerful than the AL-7F, was much more reliable. Consequently on 27th November 1961 the Soviet Council of Ministers (= government) issued a directive ordering Su-9 production to be terminated in 1962 and cutting the Su-11's production run dramatically for the benefit of the Yak-28P.

Thus OKB-51 was now facing not just further programme cuts but the daunting prospect of being closed altogether for a second time as unnecessary. Considering the disdainful attitude of the nation's political leaders towards manned combat aircraft, the chances of developing all-new aircraft were

A late-production Su-9 interceptor armed with four RS-2-US missiles.

'10 Red', the tenth production Su-11, armed with two R-8M medium-range AAMs.

The P-1 experimental interceptor – the first Sukhoi jet with lateral air intakes.

close to zero; all the OKB could do was modernise existing designs, and then only if state-of-the-art missile armament was integrated did these plans have any chance of success.

This was the situation in which the Sukhoi OKB began development of the **T-58** single-engined interceptor – the first aircraft to have this designation. To win support at the top echelon the project was disguised as a 'further upgrade of the Su-11'. The OKB-339 avionics design bureau offered two alternative radars for the T-58 – the *Oryol-2* (Eagle-2), an upgrade of the Su-11's RP-11 Oryol radar, and the brand-new Vikhr' (Whirlwind). However, both radars were too bulky to fit inside the shock cone of an axisymmetrical nose air intake as used on the Su-11; hence the radar occupied the entire fuselage nose, the AL-7F-2 engine breathing through two-dimensional (rectangular-section) lateral intakes with vertical airflow control ramps – a design that was not yet fully explored in the Soviet Union at the time. By then OKB-51 already had some experience with lateral air intakes, having used them on the P-1 delta-wing two-seat experimental interceptor of 1956 and the T-49 experimental interceptor of 1959 (basically a modified Su-11).

The conical radome mated with a basically cylindrical forward fuselage that was flattened from the sides in the cockpit area and flanked by the air intakes which blended smoothly into a centre fuselage of basically cylindrical shape. The rear fuselage structure, wings, tail unit and landing gear were identical to those of the Su-11. The armament consisted of two air-to-air missiles (AAMs).

Prototype construction began at MMZ No.51 in July 1960. But then the military started demanding ever-higher performance; the T-58 was required to have all-aspect engagement capability (that is, in both pursuit and head-on mode) against targets flying at up to 27,000 m (88,580 ft) and 2,500 km/h (1,550 mph). In keeping with a Council of Ministers directive drafted in November 1960 the interceptor was to be equipped with the Vikhr'-P radar and the *Polyot* (Flight) ground controlled intercept (GCI) system and armed with two K-40 AAMs; the choice of the missile type was dictated by the military who also envisaged this weapon for Mikoyan's new interceptors. The aircraft was allocated the service designation **Su-15**. It was the core of an aerial intercept weapons system provisionally designated T-3-8M2 because the aircraft would be armed with K-8M2 AAMs (a product of Matus R. Bisnovat's OKB-4) pending availability of the intended K-40s; the K-8M2 was a refined version of the K-8M which, unlike its precursor, had all-aspect engagement capability.

Still, time passed but the promised K-40 missile was nowhere in sight. OKB-51 continued development of the interceptor with the alternative Oryol-2 radar and K-8M2 AAMs, but the work on the T-58 project was suspended in the summer of 1961. 'For want of a missile the fighter was lost'? Well, not exactly.

Acknowledgements
The book is illustrated with photos by Yefim Gordon, the late Sergey Skrynnikov, Sergey Popsuyevich, ITAR-TASS, the Novosti Press Agency (APN), as well as from the archive of the Sukhoi Company JSC, the M. M. Gromov Flight Research Institute (LII), the personal archive of Yefim Gordon and from the following web sources: www.karopka.ru, www.scalemodels.ru, www.rumodelism.com, www.scalemates.com, www.modellversium.de, www.modelclub.gr, www.hyperscale.com, www.aircraftresourcecenter.com, www.ipmsusa3.org, www.arcair.com, www.jonbryon.com, www.ebay.com. Line drawings by the late Vladimir Klimov. Colour drawings by Andrey Yurgenson.

The Su-15 is born

As an insurance policy in case the single-engined T-58 was rejected, in late 1960 OKB-51 prepared a new version of the project envisaging installation of two 7,200-kgp (15,870-lbst) R21F-300 axial-flow afterburning turbojets side by side in the rear fuselage; this engine was developed by OKB-300's new Chief Designer Nikolay G. Metskhvarishvili. The PVO General Headquarters insisted that the twin-engined version (likewise officially designated Su-15) be equipped with the Vikhr'-P radar and armed with two K-40 AAMs, even though using the Oryol-2 radar and K-8M2 AAMs would allow the interceptor to enter service much sooner.

The general arrangement and internal layout was finalised in 1961. The powerplant was changed at this stage. Firstly, the single-engined version was indeed rejected, the customer expressly demanding twin-engine reliability; secondly, the R21F-300 turned out to have serious design flaws and was abandoned in 1962. This prompted the Sukhoi OKB to select the proven R11F2-300 to power the T-58. Accommodating the two R11F2-300s in the rear fuselage presented no problem; OKB-51 already had some experience with a similar engine installation on the T-5 development aircraft of 1958 (basically a modified Su-9).

Contrary to normal Soviet practice, no project chief was assigned to the T-58D until the mid-1960s; General Designer Pavel O. Sukhoi resolved the key issues related to the interceptor's design, while the problems arising in the course of day-by-day work were handled by his deputy Yevgeniy A. Ivanov.

In the course of 1961 the OKB completed the detail design of the interceptor whose in-house designation was now amended to **T-58D**, the D suffix standing for either ***dvigateli*** (engines, as a reference to the new twin-engine powerplant) or ***dorabotannyy*** (modified). The prototype and the static test airframe were converted from the unfinished airframes of the cancelled single-engine T-58 *sans suffixe*; stock Su-11 subassemblies modified to match the new area-ruled fuselage (which was the only major component designed from scratch) were used to save time. Another important change occurred at this point; since the K-40 AAM had been selected as the main weapon for the Mikoyan Ye-155P heavy interceptor (the future MiG-25P *Foxbat-A*), it was agreed that the T-58D (Su-15) would be armed with two K-8M2 missiles in semi-active radar homing (SARH) and infra-red homing versions, provided that a further improved version of the Oryol-2 radar designated **Sobol'** (Sable) was used.

Since the T-58D was a lot heavier than the Su-11 while having the same wing area, field performance would clearly deteriorate. To compensate for this the designers decided to use blown flaps instead of the Su-11's area-increasing Fowler flaps, the air for these being bled from the engine compressors. The landing gear was reinforced to cater for the higher weight.

The general belief was that the T-58D would have to deal primarily with single low-manoeuvrability targets flying at altitudes of 2,000-24,000 m (6,560-78,740 ft) and speeds up to 2,500 km/h (1,550 mph). With-

The T-5 development aircraft – a twin-engine experimental derivative of the Su-9. This view shows the wider rear fuselage accommodating two R11F-300 turbojets side by side.

The T58D-1 (the first prototype Su-15) nearing completion in the prototype assembly shop of the Sukhoi OKB in early 1962.

The port air intake of the partially completed T58D-1, showing the airflow control ramp actuator.

The T58D-1 with the rear fuselage detached, showing the installation of the starboard R11F2S-300 engine and its afterburner; one of the nozzle actuators is clearly visible.

out a significant advantage in speed the interceptor stood no chance of destroying such targets in pursuit mode; hence high-speed targets were to be intercepted in head-on mode, and both tactics would be used against slower aircraft. The technique of intercepting targets flying at altitudes beyond the fighter's service ceiling had been perfected with the Su-11. It involved climbing to a so-called base altitude where the fighter would be guided towards the target by GCI centres, subsequently tracking it with its own radar; after coming within missile launch range the fighter would pull up, firing the missiles in a zoom climb. The minimum missile launch altitude was restricted by the performance of the radar which lacked 'look-down/shoot-down' capability.

To automate the intercept procedure insofar as possible the T-58D featured a purpose-built automatic flight control system (AFCS), subsequently designated SAU-58 (*sistema avtomaticheskovo oopravleniya* – automatic control system), which included heading adjustment command modules and pre-programmed optimum climb profiles. In the course of GCI guidance and the actual intercept the pilot could choose between three control modes – manual, semi-automatic (flight director mode) and fully automatic.

The obligatory in-house review of the advanced development project and the mock-up review commission (a project analysis by the customer for the purpose of eliminating grave shortcomings at an early stage) were dispensed with, since the T-58D was considered to be merely an upgraded version of the Su-11. Designated T58D-1 and wearing the very appropriate (and unusual) tactical code '58-1 Red', the first prototype of the new interceptor was completed in the first quarter of 1962, making its maiden flight on 30th May 1962 with Sukhoi OKB chief test pilot Vladimir S. Il'yushin at the controls. By the end of the year it had made 56 flights under the manufacturer's flight test programme, largely confirming the expectations of its creators.

On 17th September 1962 GKAT issued an order specifying that the T-58D be equipped with a new *Smerch-AS* (Tornado-AS) fire control radar; thus the Ye-155P, Tu-128 and T-58D would have radar commonality. However, a change of radar would entail a redesign of the T-58D's forward fuselage; also, no prototype radar was available for installation. Besides, the new Sukhoi interceptor flew well and could be rapidly put into production; the integration of a new radar could delay production and service entry for years. The Sukhoi OKB pushed for a decision to use the upgraded Oryol radar on the T-58D – and succeeded in making its point not only to GKAT's top executives but also to the Commanders-in-Chief of the VVS and the PVO. On 13th March 1963 Council of Ministers Vice-Chairman Dmitriy F. Ustinov gave the go-ahead to use the Oryol-D58 radar, the D standing for *dora**bot**annyy* (modified) – with the proviso that the Smerch radar would be integrated eventually. At the same time the military curbed their appetites a little, reducing the maximum target speed to 2,000 km/h (1,240 mph) and the interception altitude to 23,000 m (75,460 ft). Thus the OKB managed to buy some time.

The second and third prototypes – the T58D-2 (coded '32 Red') and T58D-3 ('33 Blue'), which joined the programme in May and October 1963, were equipped with the Oryol-D58 radar. New design changes were progressively introduced. In particular, a more pointed radome was fitted to reduce drag; a 400-mm (15¾ in) 'plug' was inserted at the base of the fin to improve directional stability, conveniently providing accommodation for the brake parachute container relocated from the rear fuselage underside. State acceptance trials of the T58D-2 began in Zhukovskiy on 5th August 1963. Since development of the SAU-58 AFCS was running behind schedule, it was decided to hold a separate trials programme when the system became available. A State commission chaired by Air Marshal Yevgeniy Ya. Savitskiy, C-in-C of the PVO's fighter aviation, was appointed for the trials, which were performed by the Air Force's project test pilots Stal' A. Lavrent'yev, Leonid A. Peterin and Vadim I. Petrov plus Sukhoi OKB pilot Vladimir S. Il'yushin.

The pilots were generally pleased with the aircraft's handling but pointed out that aileron authority decreased at low speeds, complicating crosswind landings. Another criticism was that the engines tended to run roughly during certain vigorous manoeuvres with a sideslip. These shortcomings resulted from the aircraft's layout; thus, the unstable engine operation during manoeuvres with a sideslip was caused by the lateral air intakes – the intake on the opposite side to the direction of the sideslip was blanked off by the fuselage. The take-off and landing speeds were rather high, too; this and the unimpressive field performance was due to the fact that the boundary layer control system (BLCS) was inactive – the version of the R11F2-300 engine featuring bleed valves for the blown flaps was still unavailable. The acceleration parameters had deteriorated as compared to the Su-11, but one has to remember that the T-58D was heavier; not only the all-up weight but also the empty weight was 1.5 tons (3,306 lb) higher – 10,060 kg (22,180 lb) versus 8,560 kg (18,870 lb).

Stage B of the state acceptance trials at the Soviet Air Force's Red Banner State Research Institute named after Valeriy P. Chkalov (GK NII VVS – *Gosoo**dars**tvennyy Krasnoznamyonnyy na**ooch**no-iss**ledo**va-tel'skiy insti**toot** vo**yen**no-voz**doosh**nykh seel*) lasted from August 1963 to June 1964, involving live missile launches against real targets. Actually the trials involved not just the aircraft itself but the entire Su-15-98 aerial intercept weapons, which comprised the Su-15 interceptor, the fire control radar and two updated K-8M1P AAMs which received the new designation K-98 (in SARH

The T58D-1 with the very appropriate (and very non-standard) tactical code '58-1 Red' during initial flight tests with two K-8M1P AAMs and two drop tanks. Note the initial short radome.

and IR-homing versions). The system worked with the ***Vozdukh-1*** (Air-1) GCI system.

The trials of the Su-15-98 weapons system progressed smoothly like never before, the customer voicing almost no criticisms. By early December 1963 a total of 87 flights had been made under the trials programme. The aircraft's flight performance and the operation of its systems and equipment had been fully explored, the performance figures largely matching the manufacturer's estimates. Being a highly experienced pilot who had seen a lot of combat in the Great Patriotic War of 1941-45, and a representative of the customer into the bargain, Air Marshal Savitskiy would be ill advised to squander praise on aircraft which were not worthy of it; thus his positive appraisal of the T-58D testifies to the undoubted success achieved by the Sukhoi OKB.

The first K-98 missiles were delivered to GK NII VVS in early 1964, allowing the T-58D's armament trials to begin; these included missile attacks in head-on mode. The missile basically met its specifications, except for the inability to score guaranteed 'kills' against high-speed targets because the proximity fuse could not detonate the warhead in time at high closing speeds. The verdict was that in a head-on attack the interceptor's missile armament enabled it to destroy targets doing up to 1,200 km/h (745 mph). The combat radius and effective range were shorter than expected, ferry range at optimum altitude with two drop tanks being only 1,260 km (780 miles) instead of the required 2,100 km (1,300 miles). Hence the State commission recommended that the internal fuel capacity be increased. The Sukhoi OKB decided to do so by eliminating the 'waist' of the area-ruled fuselage. Within a short time the T58D-1 was modified in the first quarter of 1964 by riveting on a sort of 'corset' over the narrow centre fuselage so that the latter had constant width; this increased the internal

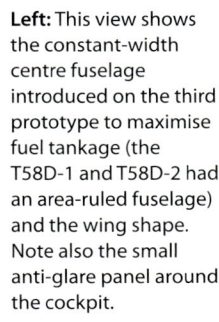

Left: This view shows the constant-width centre fuselage introduced on the third prototype to maximise fuel tankage (the T58D-1 and T58D-2 had an area-ruled fuselage) and the wing shape. Note also the small anti-glare panel around the cockpit.

Opposite page: '33 Blue', the third prototype Su-15 (T58D-3), shows off the longer, more pointed radome. Note the non-standard pylon-mounted antenna pod under the nose and the five stars under the cockpit to mark test launches of missiles.

These views show the T58D-3's rear end with the taller vertical tail and the brake parachute container relocated to the base of the fin. The red-painted K-98 AAMs are dummy versions.

volume sufficiently to provide an internal fuel capacity of 6,860 litres (1,509 Imp gal) – more than the total pre-modification capacity **with** drop tanks. Additionally, to improve stability and handling the aileron travel was increased from 15° to 18°30' and the air intake ramp adjustment time was reduced from 12 to 5-6 seconds. A special test programme at GK NII VVS on 2nd-16th June 1964 showed good results, and the modifications were recommended for introduction.

The state acceptance trials were officially completed on 25th June 1964. The final report said that the new aircraft offered significant advantages over the production Su-11, especially as far as head-on engagement capability was concerned. Other strengths included greater flight safety, longer target detection/lock-on range, a lower minimum operational altitude and the new radar's better electronic countermeasures (ECM) resistance. The report went on to say that the T-58 could intercept targets at medium and high altitudes round the clock and in adverse weather, the aircraft was easy to fly and could be mastered by the average pilot; the aircraft was recommended for service entry. It was recommended to explore the possibility of operating the Su-15 from unpaved runways after the wheels had been replaced with skids.

On 30th April 1965 the Council of Ministers issued a directive formally including the Su-15-98 aerial intercept weapons system into the PVO inventory. Thus the aircraft officially received the service designation Su-15 (ie, it was recognised as a new type, not a 'modernised Su-11'), while the K-98 missile was redesignated R-98. The directive required the Novosibirsk aircraft factory No.153 to launch series production of the new interceptor in early 1966. In the course of 1965 the plant geared up to build the Su-15, which bore the in-house code *iz**del**iye* (product) 37; later, when the Su-15T/Su-15TM entered production, the original model was sometimes referred to in paperwork as '*izdeliye* 37 Series D' in order to distinguish it from the new version, or '*izdeliye* 37 Series M'. At this time N. P. Polenov was appointed Su-15 project chief at OKB-51.

Building the Su-15 did not involve a complete change of manufacturing technology, since the aircraft had considerable structural commonality with the Su-9 and Su-11 which the factory had built earlier; it also had a lot in common with both the Su-11 and the Yak-28P as regards systems and equipment. Yet the first pre-production example (construction number 0015301 – i.e., Batch 00, plant No.153, 01st aircraft in the batch) took a long time to complete; it was rolled out at Novosibirsk-Yel'tsovka on 21st February 1966, several months late. On 6th March the aircraft made its first flight with factory test pilot Ivan F. Sorokin at the controls. The second pre-production Su-15 (c/n 0015302) was completed in June 1966, joining the first aircraft on 21st July.

In February 1965 the first prototype had been refitted with double-delta wings of increased area and 720 mm (2 ft 4⅜ in) longer span, the leading-edge sweep on the outer portions being reduced from 60° to 45°; the purpose of this modification was to increase aileron efficiency at low speeds. However, the Novosibirsk factory refused to build the fighter with the new wings on the pretext that no verdict from GK NII VVS existed on the soundness of this design. The real reason was more down-to-earth but nonetheless plausible: the jigs and tooling for the Su-15's original delta wings had already been manufactured; modifying them would mean further delays in the production schedule, which was already in jeopardy.

Full-scale production of the initial Su-15 *sans suffixe* gained momentum from mid-1966 onwards, peaking in 1969 when 165 examples were built; the last machine was completed in 1971. The production version had an internal fuel capacity of 6,860 litres (like the post-modification T58D-3) but the number of fuselage tanks was reduced to three. Production aircraft were powered by R11F2S-300 (*izdeliye* 37F2S) engines rated at 3,900 kgp (8,600 lbst) dry and 6,175 kgp (13,610 lbst) reheat. The SAU-58 AFCS took some time coming, commencing trials only in 1968; the trials revealed the need for a major redesign and the introduction of the AFCS was postponed until the aircraft's next upgrade. As a result, the aircraft had neither the AFCS nor the simpler AP-28 autopilot fitted to the prototypes; the road to hell is paved with good intentions! Worse, the production Su-15 did not even have dampers in any of the control circuits, the ARZ-1 artificial-feel unit in the tailplane control circuit being the only stability augmentation system. Curiously, the military never demanded changes in this area, as the stability and handling characteristics of production aircraft during pre-delivery tests appeared adequate.

The production version featured an improved version of the air intake control system designated UVD-58M and the Sukhoi KS-4 ejection seat permitting safe ejection throughout the flight envelope, providing the speed was above 140 km/h (87 mph). The avionics fit of early-production Su-15s included an RSIU-5 (R-802V) VHF communications radio, an MRP-56P marker beacon receiver, an RV-UM low-range radio altimeter, an ARK-10 ADF, SOD-57M distance measuring equipment, an ARL-S *Lazoor'* (Prussian Blue) GCI command link receiver, an SRZO-2M identification friend-or-foe (IFF) interrogator/transponder, a *Sirena-2* (Siren) radar warning receiver, a KSI-5 compass system and an AGD-1 artificial horizon. In its production form the Oryol-D58 radar was officially designated RP-15 (**rah***diopritsel* – 'radio sight', the Soviet term for fire control radars). Increasing the radar's ECM resistance and perfecting the R-98 missile's proximity fuse was the responsibility of other bureaux.

Head-on view of the first pre-production Novosibirsk-built Su-15 (c/n 0015301), showing the distinctively-canted air intakes.

Two more views of the same aircraft at Zhukovskiy. Note that the tactical code '34 Red' continues the sequence of the prototypes.

Three-quarters rear view of Su-15 c/n 0015301, showing the prominent dorsal air intake scoops for cooling the engine nozzles.

This aspect of the first pre-production aircraft shows the enlarged anti-glare panel on the nose and the additional anti-glare panels on the air intakes.

Su-15 c/n 0015301 in later days as '01 Red', seen parked at the premises of the Sukhoi OKB in Moscow. Note the striped 'Danger, air intake' stencils.

'67 Red', the second production Su-15 (c/n 0115302), with photo calibration markings applied to the forward and rear fuselage. Note that the aircraft is depicted on an unpaved runway. This machine was used to investigate a problem associated with abruptly increasing control forces.

This, too, proved to be a protracted affair and the trials of the improved RP-15M (Oryol-D58M) radar were not completed until 1967 when the Su-15 had become operational. New-build Su-15s were henceforth completed with the modified radar, while previously built examples were upgraded *in situ* to the new standard.

Between 7th July 1967 and 25th September 1968 GK NII VVS held check-up trials of a Batch 2 Su-15 – with disappointing results. There were two main problems. The first of these lay with the engines. The R11F2S-300 turbojet was produced by two factories, No.500 in Moscow and No.26 in Ufa, Bashkiria; in the course of the tests it was seen that the Moscow-built engines provided the required performance but the Ufa-built ones did not (this primarily concerned interception range in pursuit mode). On 10th June – 30th August 1969 the same aircraft underwent renewed tests with a new pair of Ufa-built engines; this time the performance improved to an acceptable level, except that interception range in pursuit mode against a target doing 1,400 km/h (870 mph) at 23,000 m (75,460 ft) was 25 km (15.5 miles) shorter than required. GK NII VVS immediately solved the problem by a bit of trickery: the next mission was flown with one drop tank instead of two to reduce drag, and the required interception range of 195 km (121 miles) was obtained.

The other nasty surprise was that in certain flight modes the stick forces would suddenly increase abruptly; the problem was traced to the hydraulic control surface actuators which were not powerful enough. The OKB and the military turned to the Flight Research Institute named after Mikhail M. Gromov (LII – **Lyot**no-is**sled**ovatel'skiy insti**toot**) for help; at the end of 1968 the institute conducted a special research programme to explore this. Concurrently the OKB investigated the problem on its own, but soon pressure of higher-priority work caused the issue to be put on hold.

The Su-15 underwent constant refinement in the course of production. The biggest number of upgrades was made in 1968 when the double-delta wings and the BLCS were introduced after all. The first shipset of R11F2SU-300 (*izdeliye* 37F2SU) engines with BLCS bleed valves was delivered to OKB-51 in early 1968, whereupon the first pre-production Su-15 was re-engined and rewinged. After a brief manufacturer's flight test programme the fighter underwent more extensive testing at GK NII VVS. The extended-chord outer wing portions had a positive effect on the fighter's stability, but the double-delta wings caused a slight reduction in supersonic performance and a reduction of the service ceiling; still, the performance figures were within the limits specified by the military. The blown flaps caused a forward shift in the aircraft's CG, which complicated the landing procedure because the stabilator travel limits were insufficient; on the other hand, approach speed was reduced by 40 km/h (25 mph). The final verdict was a thumbs-up and in 1969 the blown flaps were introduced on production Su-15s from Batch 11 onwards but the old pure delta wings were retained for a while; the first production Su-15 built with double-delta wings was c/n 1115331. From c/n 1115336 onwards all Su-15s had provisions for installing the new Gavrilov R13-300 afterburning turbojets (more about this later).

Pre-delivery tests showed that the modified wings reduced the interceptor's service ceiling by an average 400 m (1,310 ft). As a result, the PVO temporarily stopped accepting new Su-15s; Pavel O. Sukhoi's aide Yevgeniy A. Ivanov, who was responsible for the Su-15 programme, received a dressing-down from the 'head office' at GKAT. Yet, both the military and the OKB were well aware that nothing could be done to increase the service ceiling; hence an agreement was signed in which the PVO conceded that the service ceiling of the rewinged Su-15 would be 18,100 m (59,380 ft) versus 18,500 m (60,695 ft) for the original 'pure delta' version. As for the control force issue, it was resolved by installing more powerful actuators after the cranked-wing version had entered production. Tests held in 1970 with the new actuators showed that the phenomenon recurred only when the airbrakes were

Above and right: A trio of early-production (pure delta) Su-15s make a flypast at the airshow at Moscow-Domodedovo airport on 9th July 1967.

Right: One of the first production Su-15s participating in the 1967 Domodedovo airshow wore an unusual sinister overall black colour scheme with the tactical code '47 Red' applied in a non-standard typeface.

Late-production Su-15s *sans suffixe* from c/n 1115331 onwards had new double-delta wings, as illustrated by this photo.

deployed or if one of the hydraulic circuits failed. From then on identical BU-220 actuators were installed in all three control circuits on production Su-15s.

The powerplant also had its share of bugs. Though more reliable than the AL-7F, the R11F2S-300 engines powering the Su-15 were not immune against surging or flaming out at high altitude, and relighting them after a flameout was not always easy. When the fighter entered service, it turned out that the engines were prone to catching fire, while the fire warning and fire suppression systems were unreliable. In 1968-69 alone, two accidents and six incidents occurred in PVO fighter units for these reasons.

The production Su-15 *sans suffixe* made its public debut on 9th July 1967 during the grand airshow at Moscow-Domodedovo airport. After that the NATO's Air Standards Co-ordinating Committee (ASCC) allocated the reporting name *Flagon* to the new interceptor; this was changed to *Flagon-A* when new versions became known. The R-98 AAM was codenamed AA-3 *Anab*.

Some early-production Su-15s were used in various test programmes. These included '16 Blue' (c/n 0315306), a 611th IAP aircraft; on 9th July 1967 this aircraft participated in the Domodedovo airshow.

For some reason, dual-control trainer versions of Soviet fighters usually appeared much later than the baseline single-seaters, and the Su-15 was no exception. The first project studies of a two-seat version of the T-58 dated back to 1961-62 but the design work was suspended due to pressure of higher-priority programmes. Hence the Su-15 had to be mastered by service units without the benefit of a trainer. Officially development of the latter was initiated only on 30th April 1965 by a special item of the CofM directive clearing the Su-15 for service. The appropriate Ministry of Aircraft Industry (MAP – *Ministerstvo aviatsionnoy promyshlennosti*) order appeared on 20th May 1965, requiring OKB-51 to build two prototypes and a static test airframe; the aircraft was to commence state acceptance trials in the second quarter of 1967. The aircraft received the manufacturer's designation **U-58** (the prefix denoted *oochebnyy* [*samolyot*] – trainer).

The trainer had tandem cockpits for the trainee and the instructor; this necessitated a 450-mm (1 ft 5¾ in) fuselage stretch aft of the existing cockpit. The cockpits were enclosed by a common canopy similar to that of the Su-9U *Maiden*, with individual aft-hinged portions and a fixed section in between. Each cockpit featured a full set of controls and instruments and a KS-4 ejection seat. The front cockpit featured a blind flying hood, while the rear canopy portion incorporated a retractable periscope to give the instructor a measure of forward view during take-off and landing.

Since the U-58 was intended for both flight training and live weapons training, it was to retain the complete avionics and armament fit of the late-production single-seater and have similar performance. Thus the trainer was meant to carry two R-98 AAMs and feature an advanced **Korshun-58** (Kite, the bird) radar, a derivative of the Oryol-D58.

The in-house project review and the sessions of the mock-up review commission took place in October 1965. A full set of manufacturing documents for the U-58 was delivered to the OKB's Novosibirsk branch office by September 1966. By then it was obvious that the manufacture of the trainer airframes at plant No.153 was running behind schedule; the delivery date was slipping and hence the state acceptance trials deadlines would have to be moved. Also, in 1967 the military demanded that the new *Taïfoon* (Typhoon) fire control radar be integrated on the upgraded (double-delta) Su-15 instead of the Korshun radar; this meant the Taïfoon would have to be fitted to the trainer as well, causing further delays. Hence the Sukhoi OKB suggested splitting the U-58 programme into two stages to speed up progress – a suggestion gladly accepted by MAP. Stage One involved developing a simplified conversion trainer lacking radar and some other equipment items; these would be added during Stage Two to create a fully capable combat trainer.

The downgraded conversion trainer variant received the manufacturer's designation **U-58T** and the service designation **Su-15UT** (*oochebno-trenirovochnyy* – for [conversion and proficiency] training) to indicate it had no combat capability. The aircraft featured a standard navigation and communications suite as fitted to the single-seat Su-15 but lacked the latter's fire control radar, Lazoor'-M GCI command link system, radar warning receiver and missile arming/launch system modules. On the other hand, the avionics were augmented by the addition of an SPU-9 intercom (*samolyotnoye peregovornoye oostroystvo*) and an MS-61 cockpit voice recorder. The bottom line was that the trainer's avionics fit was rather basic. Like the single-seater, the Su-15UT had missile pylons but these could carry only dummy missiles. The rear cockpit encroached on the No.1 fuselage fuel tank, reducing its capacity by 900 litres (198 Imp gal); this was partly compensated by adding the 180-litre (39.6 Imp gal) No.5 fuel tank in the rear fuselage beneath the engines. All of this caused a rearward shift in the CG position, which was restored by installing ballast in the forward fuselage. Empty weight increased to 10,660 kg (23,500 lb); this, together with the reduction of the fuel capacity, entailed a substantial reduction in range.

The static test airframe of the U-58T was completed in late 1967. Designated U58T-1, the first prototype ('01 Red') followed in the summer of 1968; it had pure delta wings and the BLCS was inactive. Sukhoi OKB test pilot Yevgeniy K. Kukushev was appointed project test pilot, performing the trainer's maiden flight at Novosibirsk-Yel'tsovka on 26th August; by 16th September the aircraft had been ferried to Zhukovskiy where the manufacturer's test programme began. MAP kept pushing for the state acceptance trials to begin as soon as possible; hence as early as 2nd October the U58T-1 was ferried to GK NII VVS at Vladimirovka AB. Manufacturer's tests continued there, proceeding in parallel with the state acceptance trials until 12th December 1968.

On 15th-19th October 1968 the U58T-1 was test flown by a number of pilots from operational PVO units, including the new PVO Aviation C-in-C Lt.-Gen. Anatoliy L. Kadomtsev. The actual flights under the state acceptance trials programme did not begin until 16th November. GK NII VVS project test pilots Mikhail I. Bobrovitskiy, Gheorgiy A. Bayevskiy and Nikolay V. Rukhlyadko did most of the flying. The programme included performance testing, assessment of field performance, stability and handling; part of it was performed by OKB test pilots, as usual.

The state acceptance trials were completed on 26th February 1969. Predictably, the two-seater's performance was inferior to that of the single-seater: range had dropped to 1,390 km (863 miles) and the service ceiling had decreased to 17,700 m (58,070 ft). Stability and handling were deemed accept-

Above: '01 Red', the Su-15UT prototype (the U58T-1). Note the canopy design; the hinged canopy section over the trainee's seat is longer than the one over the instructor's seat. Though depicted with R-98T missiles, the Su-15UT could not use them for want of missile control equipment.

Left: The Su-15UT trainer was built in limited numbers alongside the combat version.

able, except that some directional instability set in above Mach 1. Generally, however, GK NII VVS stated that the Su-15UT was suitable as a pilot trainer, except for weapons training which was beyond the aircraft's capabilities.

The trials report was endorsed in the summer of 1969 and the OKB set to work eliminating the trainer's deficiencies. The cause of the instability mentioned above was clear – the vertical tail area was too small, now that the area ahead of the CG had increased. Increasing the vertical tail area by inserting another 'plug' at the root, as had been the case in the Su-15's early flight test days, was impossible for structural strength reasons. The OKB decided to try equipping the Su-15UT with ventral fins. However, a special test programme held with the modified prototype in the spring of 1970 showed that the fins did not give any major improvement; therefore the OKB and the Air Force compromised, limiting the trainer's maximum speed to Mach 1.75. Consequently the service ceiling further decreased to 16,700 m (54,790 ft).

The Su-15UT entered production at the Novosibirsk plant in 1969 under the in-house product code *izdeliye* 42; the first production aircraft (c/n 0115301) was completed in October, making its first flight on 10th December. Production examples differed from the prototype in having double-delta wings and an operational BLCS. The first five Su-15UTs were delivered to operational units in the spring of 1970; in July that year the aircraft was officially included into the inventory. In addition to first-line units, the type saw service with the PVO's Stavropol' Military Pilot College and LII's Test Pilots School in Zhukovskiy. The NATO reporting name was *Flagon-C*.

The Su-15UT remained in production until the end of 1972. The final batches featured a new R-832M Evkalipt-SMU communications radio replacing the earlier R-802V. The operating empty weight of the production model grew to 10,750 kg (23,700 lb).

Completed much later than intended due to late deliveries of the mission avionics, a single Su-15 was built as a fully capable combat trainer designated **U-58B** (*boyevoy* – combat, used attributively). Outwardly it was identifiable by the reinforced twin-wheel nose gear unit fitted to cater for the higher weight of the forward fuselage caused by the installation of the radar. Known at the Novosibirsk aircraft factory as *izdeliye* 37UB, the aircraft ('70 Blue', c/n 0003UB86) made its first flight on 24th June 1970 at the hands of factory test pilot Aleksey S. Gribachov and was ferried to Zhukovskiy on 2nd August. Originally the aircraft was powered by R11F2S-300 engines but these were replaced by R11F2SU-300s in May 1971, allowing the BLCS to be activated.

Unlike the Su-15UT, which was 'tail heavy' and required ballast in the nose, the U-58B was 'nose heavy' due to the combination of a rear cockpit and a radar; hence it was sluggish and generally disappointing in its performance. At the initiative of the OKB, with MAP's formal agreement, the U-58B's development was suspended; a while later the military also gave their consent to this decision. No more were built.

As already noted, the military were not entirely happy with the performance of the production Su-15. A mid-life upgrade was developed in due course, and in contemporary documents the aircraft was referred to as the 'Su-15 Stage II'. The requirements for this version were outlined by the aforementioned CofM directive of 30th April 1965. At that time the Sukhoi OKB was swamped with work, so it was less than overjoyed at the prospect of having to upgrade the Su-15. Hence most of the design work on the upgrade proceeded on a 'time permitting' basis due to the programme's low-priority status; actually it did not begin in earnest until early 1966, and even then for the first two years it made painfully slow progress. This was because the new radar type was not finalised for a long time.

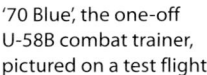
'70 Blue', the one-off U-58B combat trainer, pictured on a test flight.

In mid-1966 the OKB started detail design work on incorporating the advanced Korshun-58 radar, the SAU-58 AFCS, the RSBN-5S *Iskra* (Spark) short-range radio navigation system (**rah**dio**tekhnich**eskaya sis**te**ma **blizh**ney navi**gah**tsiï – SHORAN), a new communications radio and a skid landing gear on the Su-15. When the military belatedly discovered that the Korshun radar could not provide the required performance, the attention focused on the aforementioned Smerch radar. As a result, in October 1967, when the OKB had all but completed the project documents for the Su-15's upgrade, it was greeted by a ruling to the effect that all work on the Korshoon-58 radar be stopped and the fighter be equipped with the Taïfoon radar (a derivative of the Smerch) instead. Officially this change was documented by a ruling of the Military Industry Commission on 22nd March 1968. It was back to the drawing board.

The upgrade programme was divided into two stages. Stage One involved state acceptance trials with the existing R-98 AAMs in November 1968; during Stage Two scheduled for the third quarter of 1969 the aircraft was to be armed with the new K-98M missile (which had not yet passed its state acceptance trials either). The aircraft modified to Stage One specifications was to be designated **Su-15T**, the suffix letter referring to the Taïfoon radar, while the Stage Two aircraft would be designated Su-15TM. This gave the avionics and defence industry a respite, allowing the aircraft's new avionics and armament to be put through their paces.

The advanced development project of the Su-15T (known in-house as the T-58T) was completed by early September 1968, the in-house project review and the sessions of the mock-up review commission taking place in October. The aircraft was to be powered by new R13-300 turbojets – a derivative of the R11F-300 developed by the Ufa-based *Soyooz* (Union) engine design bureau led by Sergey A. Gavrilov. The R13-300 differed significantly from the precursor; among other things, the number of high-pressure compressor stages was increased from three to five and a second afterburner stage was added. the engine delivered 4,100 kgp (9,040 lbst) dry and 6,600 kgp (14,550 lbst) in full afterburner versus 3,900 kgp (8,600 lbst) and 6,175 kgp (13,610 lbst) respectively for the R11F2S-300; changes to the hydraulics and electrics were also envisaged.

The military demanded that the Su-15T and the Su-15TM should have a secondary strike capability that would make them usable as tactical fighters. The air-to-ground weapons options were one or two 500-kg (1,102-lb) bombs; up to four 100- or 250-kg (220- or 551-lb) bombs; one or two UB-16-57U rocket pods (*oonifitseerovannyy blok* – standardised [rocket] pod), each holding sixteen 57-mm (2.24-in) S-5 folding-fin aircraft rockets; one or two S-24 heavy unguided rockets; and two UPK-23-250 pods (*oonifitseerovannyy* **push***echnyy kon***tey***ner – standardised gun pod*) with 23-mm cannons (carried under the wings, not on the fuselage pylons). A built-in cannon was also considered.

The control system was modified by incorporating the servo drives of the SAU-58 AFCS which was finally nearing the end of its development. Apart from the usual speed/altitude/angle stabilisation and 'panic button' (automatic restoration of straight and level flight) functions, the SAU-58 was to enable automatic flight along several preset trajectories and automate the main stages of the intercept. Additionally, the AFCS was to enable automatic low-level terrain-following flight; this totally new feature was not meant for air defence penetration, of course – it resulted from the new requirement that the upgraded Su-15 was to be capable of intercepting targets flying at altitudes down to 500 m (1,640 ft). Since the Taïfoon radar still lacked 'look-down/shoot-down' capability, the intention was to 'paint' the targets from below, flying at less than 500 m. This meant flying dangerously close to the terrain at speeds close to 1,000 km/h (620 mph), working the radar all the while! Hence the OKB opted for a simpler version of the AFCS using the radio altimeter, not the radar, as the primary source of data; this made flight level stabilisation possible but the feature was usable only over flat or very moderately hilly terrain.

The 'Su-15 Stage II' was to feature a new R-832M Evkalipt-SM communications radio replacing the earlier R-802V, a Pion-GT (Peony, pronounced *pee* **on**) antenna/feeder system, built-in test equipment (BITE) and an RSBN-5S Iskra-K SHORAN. The latter enabled semi-automatic landing approach down to 50-60 m (165-200 ft), improving the aircraft's all-weather capability considerably.

An early-production (pure-delta) Su-15 (c/n 0515348) was converted as the Su-15T prototype in late 1968; for security reasons this aircraft was referred to at the OKB as 'aircraft 0006'. With the Air Force's consent the prototype entered test with an incomplete avionics fit; nor were the R13-300 engines, skid landing gear and strike armament fitted (the prototype was powered by R11F2S-300s). On the other hand, the aircraft was refitted with double-delta wings, the Taïfoon radar and the SAU-58 AFCS; a twin-wheel nose gear unit was fitted to cater for the heavier radar.

According to some documents, the Su-15T made its maiden flight on 27th January 1969 at the hands of project test pilot Vladimir A. Krechetov; however, Krechetov's log book says 31st January. To expedite the state acceptance trials, the manufacturer's tests were suspended after only eight flights and on 6th March 1969 the Su-15T prototype arrived at GK NII VVS. The acceptance procedure dragged on until the end of May; meanwhile the OKB carried on with the manufacturer's tests. PVO Aviation Vice-Commander Maj.-Gen. Fyodor I. Smetanin

These views of Su-15T '37 Red', which became a ground instructional airframe after retirement, illustrate the combination of the old conical radome and the new tall twin-wheel nose gear unit characteristic of this interim version.

chaired the State commission, while the flight test team included GK NII VVS pilots Stal' I. Lavrent'yev, Eduard M. Kolkov, Vadim I. Petrov and Stepan A. Mikoyan, as well as OKB test pilot Vladimir A. Krechetov. The trials proceeded in two stages, the first of which was to be completed in the first quarter of 1970.

The state acceptance trials of the Su-15T were a far cry from those of the original T-58D, proceeding slowly and laboriously. The main objective of the trials was to assess the interceptor's combat capabilities with the new radar. Here there would appear to be no pitfalls, since fire control radars broadly similar to the Taïfoon had been verified on the production Tu-128 heavy interceptor and the Ye-155P which had passed its trials successfully. Nevertheless the Taifoon radar turned out to be rather troublesome. By the end of the year the prototype had made 64 flights but only 40 of them counted, the other 24 missions apparently being aborted due to malfunctions (or being training flights and the like). To avoid delaying the Su-15T's production entry GK NII VVS decided to issue a so-called preliminary report; in February 1970 Stage A of the trials was discontinued altogether. The report said that only 63 of the 87 flights made had proceeded in accordance with the plan of the trials, the remainder being devoted to perfecting the radar and the AFCS. Eventually the OKB managed to get the interceptor's principal systems up to scratch and the aircraft was ready for Stage B of the trials.

The state acceptance trials ended in mid-June 1970; Stage B was not completed in full either, as the MAP and Air Force top brass demanded the beginning of the Su-15TM's state acceptance trials pronto (this was exactly what the second prototype was intended for). In the course of Stage B the two aircraft made 58 and 14 flights respectively. The final report of the trials said that generally the weapons system met the specifications, even though the new equipment, first and foremost the Taïfoon radar (no pun intended), proved rather unreliable.

The Novosibirsk aircraft factory was to complete the first 20 production Su-15Ts in the second half of 1969; yet the first production example of the new model (initially referred to as *izdeliye* 37M or *izdeliye* 37 Srs

M) was flown by the factory's CTP Vladimir T. Vylomov only on 20th December 1970. As the trials of the more advanced Su-15TM progressed, the interest of the military in the interim Su-15T waned, and the production run was limited to a mere 20 aircraft (c/ns 0115301 through 0115310 and 0215301 through 0215310); confusingly, the new version had a separate batch numbering sequence. Later, when the Su-15TM (likewise designated *izdeliye* 37M) entered production, the Su-15T's product code was changed to *izdeliye* 38 to discern it from the newer model.

The avionics suite of the production Su-15T included a Taïfoon radar, the SAU-58 AFCS, an R-832M Evkalipt-SM radio, an RSBN-5S Iskra-K SHORAN, an ARL-SM (Lazoor'-SM) GCI command link receiver, an ARK-10 ADF, an RV-5 low-range radio altimeter, an SPO-10 Sirena-3 radar warning receiver and a Pion-GT antenna/feeder system. The latter's antennas were mounted on the air data boom at the tip of the radome and above the brake parachute container (which required the rudder to be cut away at the base). The nose gear unit featured twin KN-9 non-braking wheels. The production Su-15T was powered by the old R11F2SU-300 engines.

On 25th January 1971 GK NII VVS commenced check-up trials of the first and 15th production Su-15Ts with a view to checking the operation of the armament. The programme was completed within a month, and the results were disheartening: the aircraft proved incapable of intercepting low-flying targets. In five test flights at an altitude of 500 m (1,640 ft) the radar tracked the target on two occasions only, and then at very close range – a matter of 3 km (1.86 miles), which was totally unsatisfactory. Worse, neither could the radar reliably acquire and track the target at higher altitudes almost throughout the permitted missile launch range. Furthermore, poor electromagnetic compatibility (EMC) of various avionics was discovered, landing with the BLCS switched on was complicated due to the limited tailplane travel and so on.

As a result, the delivery of the production Su-15Ts dragged on for more than a year due to the need to rectify these defects. The aircraft did not become operational until the summer of 1972. Most of them were eventually transferred from active duty to the Stavropol' Military Pilot College, serving on as trainers. A few were transferred to MAP for use in various R&D programmes. The Su-15T's NATO reporting name was *Flagon-E*.

Here we have to go back a bit. In December 1969 the Sukhoi OKB completed the second T-58T prototype converted from the fifth production Su-15 (c/n 0115305). This aircraft was actually not a Su-15T but the **Su-15TM** prototype and featured an almost complete avionics and equipment fit, including R13-300 engines and an upgraded Taïfoon-M radar compatible with K-98M missiles instead of the basic Taïfoon. Since the state acceptance trials of the first prototype Su-15T were running late, it was decided that the second prototype should join the programme at this stage. Due to development problems with a number of systems it was more than three months before the second prototype could enter flight test; the Su-15TM first flew on 7th April 1970 with Vladimir A. Krechetov at the controls. Four days later it was ferried to GK NII VVS at Vladimirovka AB to join the state acceptance trials programme. The aircraft was flown by test pilots Eduard M. Kolkov, Valeriy V. Migoonov, Nikolay A. Mostovoy and Stal' I. Lavrent'yev.

Stage A comprising 40 flights was to verify the operation of the aircraft's principal systems. Progress was terribly slow, only six flights 'for the record' being made by the end of 1970. On 3rd February 1971 the original Su-15T prototype, or 'aircraft 0006' (c/n 0515348), which had been refitted with an upgraded Taïfoon-M radar in the meantime, was added to the first aircraft in the hope of speeding up the trials, but little good did it do. During the first three months of the year the two aircraft made a mere 42 flights between them, only three of these 42 missions being accepted 'for the record'. Neither aircraft had the widened air intakes envisaged for the production version which had been tested successfully on Su-15 c/n 1315340 (see experimental versions section below). A third prototype (c/n 0115309) joined the trials programme in 1971, making 19 test flights with the objective of exploring the structural strength limits.

Meanwhile, plant No.153 was tooling up to build the Su-15T; therefore, as already mentioned, GK NII VVS decided to issue a so-called preliminary report on the trials results in order to avoid holding up production of the modernised fighter. To this end a special test schedule was drawn up, flights under this programme commencing on 20th May 1971. By the end of June the prototypes had made a total of 123 flights under Stage A of the state acceptance trials. Still, radar operation and the guidance of the radar-homing AAMs at low altitudes was unstable, and the Air Force called a halt to the testing of 'aircraft 0006' for the time being. The other prototype kept flying for a while, but on 17th June 1971 a fire broke out just as the aircraft was taxying out for take-off; test pilot Valeriy I. Mostovoy vacated the cockpit without even taking time to shut down the engines. The fire, which had been caused by a malfunction in the oxygen system, inflicted heavy damage on the aircraft which, though not a total loss, could only be repaired by the manufacturer in Novosibirsk.

Thus the state acceptance trials of the Su-15TM effectively stopped altogether; they resumed on 26th August when 'aircraft 0006' reentered flight test after the Taifoon-M radar had been updated. In December it became clear that the rebuild of the first prototype was taking longer than expected and 'aircraft 0006' would have to shoulder the

This page:
'74 Blue' (c/n 0315302), the first production Su-15TM, in its ultimate guise, showing the new ogival radome and the additional inboard pylons carrying R-60M short-range AAMs (inert rounds in this case). The many access panels give the aircraft a patchwork appearance.

Opposite page:
The second production Su-15TM, '75 Blue' (c/n 0315303), seen during state acceptance trials at GK NII VVS. The inboard pylons have yet to be fitted and the cooling air intakes on top of the rear fuselage are still there (compare this to '74 Blue', which had them removed).

'76 Blue' (c/n 1015307), an early production Su-15TM, was used in 1975 to test the aircraft for suitability for the strike role. The aircraft carries UB-32 rocket pods on the wing pylons and UPK-23-250 cannon pods on the fuselage pylons.

Here the same aircraft is seen with a ZB-500GD napalm container on the starboard wing pylon.

remainder of the state acceptance trials. The OKB and GK NII VVS attempted to widen the scope of the trials work by using the first two production Su-15TMs, '74 Blue' (c/n 0315302) and '75 Blue' (c/n 0315303), which fully conformed to the 'Su-15 Stage II upgrade' standard as far as both airframe and avionics were concerned. The two aircraft arrived at Vladimirovka AB on 15th December; however, they had been handed over to the Air Force in such haste that the obligatory pre-delivery tests had not been performed and these had to be done at GK NII VVS instead of the factory. As a result, the two production Su-15TMs did not actually join the programme until March 1972. By then the rebuilt first prototype had returned to Vladimirovka AB, but it was a case of too little, too late – Stage A of the trials ended on 31st March.

The report on the results of Stage A said that the four participating aircraft had made 289 flights between them; of these, 81 flights were made under the Stage A programme proper and 91 under the 'preliminary report' test schedule. Nearly all performance targets had been met, with the exception of the weapons system's low-altitude performance. The maximum speed of the target being intercepted and the radar's target detection range were shorter than expected. The strong points noted by the pilots included the presence of the SAU-58 automatic flight control system and a short-range radio navigation system, both of which facilitated flying and landing, and, most importantly, the Su-15TM's greatly enhance combat potential as compared to the in-service Su-15 *sans suffixe*. The insufficient tailplane travel was cited as the main shortcoming. Eventually the strengths outweighed the weaknesses and the Su-15TM was recommended for production.

Stage B of the trials began on 17th April 1972. The greater part of it was to be performed using the first two production Su-15TMs which were equipped with production Taïfoon-M radars manufactured by LNPO Leninets, the other two aircraft being set aside for special test programmes. The radars proved fairly reliable; also, they featured a new module increasing the radar set's ECM resistance. At this stage the OKB had to tackle the complex task of increasing intercept efficiency at low altitudes (the aircraft did not meet the Air Force's requirements in this respect, and only one of five such missions performed by the end of April had been successful). A new GCI guidance algorithm had to be developed, and the result was felt immediately; 12 of the 16 low-altitude intercept missions that followed were successful, albeit they were flown over level terrain.

Stage B ended in April 1973, by which time a third production Su-15TM (c/n 0315304) had joined the test fleet. All in all, the five aircraft made 256 flights, including 143 as part of the state acceptance trials programme and 103 under various special test

programmes. 46 K-98M missiles were fired at Lavochkin La-17M, Mikoyan/Gurevich M-17 (MiG-17M), Il'yushin M-28 (Il-28M) and Tupolev M-16 (Tu-16M) target drones, some of which featured active and passive ECM gear, as well as at RM-8 and PRM-1 paradroppable targets and KRM high-speed targets based on cruise missiles. Three of the target aircraft received direct hits and three others were destroyed by proximity detonation.

The final report on the trials results did not point out any major shortcomings; on the other hand, it effectively outlined a plan for the further upgrade of the Su-15TM. Among other things, in the future the aircraft was to be re-engined with the new Gavrilov R25-300 afterburning turbojets (see Su-15*bis* below). New wings and stabilators of increased area were to be incorporated, an automatic lateral stability augmentation system and a trim mechanism were to be provided.

The Su-15TM had entered production back in October 1971, superseding the Su-15T on the Novosibirsk production line. Since the two models had a lot in common, the Su-15TM's batch numbering continued that of the Su-15T (i.e., Su-15TM production started with batch 3). The in-house product code (*izdeliye* 37M or *izdeliye* 37 Srs M) remained unaltered. This was the last of the Su-15's production versions, the final 'TMs pertaining to batch 14 rolling off the production line in late 1975. Concurrently plant No.153 produced the Su-24 tactical bomber, hence the Su-15TM's production rates were

Su-15TM '31 Blue' (c/n 1415331) was used in various test programmes. Note the numerous maintenance stencils in red.

Here the same aircraft is seen with an improbable mix of ordnance during tests for suitability for the strike role. The wing pylons carry R-98 and R-60M AAMs, while the fuselage pylons carry two 250-kg (550-lb) FAB-250 high-explosive bombs.

not particularly high, peaking at 110 aircraft per year.

The upgraded Su-15-98M aerial intercept weapons system was officially included into the inventory by a Council of Ministers directive dated 21st January 1975. It enabled manually- or automatically-controlled interception of single targets flying at altitudes of 500-24,000 m (1,640-78,740 ft) and speeds up to 1,600 km/h (990 mph) in pursuit mode and targets flying at altitudes of 2,000-21,000 m (6,560-68,900 ft) and speeds up to 2,500 km/h (1,550 mph) in head-on mode. The weapons control system had enhanced ECM resistance; on the down side, the service ceiling had decreased from 18,500 m (60,690 ft) – the figure obtained in the course of the state acceptance trials – to 17,970 m (58,960 ft). After service entry the Taïfoon-M radar received the official designation RP-26, while the K-98M missile was renamed R-98M.

The first lot of production Su-15TMs was handed over to the military in the spring of 1972. Apart from the radar, production Su-15TMs were virtually identical to the Su-15T as regards avionics and equipment. Empty weight increased to 10,870 kg (23,960 lb) versus 10,220 kg (22,530 lb) for the Su-15 *sans suffixe*; the fuel capacity was 6,775 litres (1,490.5 Imp gal) and the fuel load 5,550 kg (12,235 lb).

One of the deficiencies the Sukhoi OKB and LNPO Leninets had to rectify together was the clutter on the radar display arising from internal reflections of the radar pulse inside the radome. The avionics house suggested fitting an ogival radome to cure the problem. After a series of tests on Su-15 c/n 1315340 at LII in June 1972 involving several differently shaped radomes, such ogival radomes were fitted to the first two production Su-15TMs (c/ns 0315302 and 0315303) which underwent additional tests. The latter showed that the annoying echoes had vanished but the service ceiling, base altitude, effective range and interception range had decreased somewhat due to the extra drag created by the new radome. (Contrary to popular belief, the ogival radome was not a measure aimed at improving the Su-15TM's aerodynamics – in fact, it made them worse.)

As part of the same tests, Su-15TM '31 Blue' was operated experimentally from an unpaved runway simulating an *ad hoc* tactical airstrip.

Upon completion of the check-up trials at GK NII VVS in the autumn of 1973 the new radome was cleared for production. By then, however, a considerable number of Su-15TMs had been built with the old 'pencil nose'; new-build 'TMs received the ogival radome from batch 8 onwards and the previously manufactured aircraft were progressively updated in service. The Su-15TM received the NATO reporting name *Flagon-F*.

The service tests of the new variant were held by one of the PVO's first-line units between 1st February 1975 and 20th July 1978 with good results. Like most aircraft, the aircraft was not immune to accidents; the first total loss was on 7th February 1973 when Su-15TM c/n 0815320 suffered a critical failure during pre-delivery tests in Novosibirsk, forcing factory test pilot I. M. Gorlach to eject.

In the course of production the Su-15TM was constantly updated. In particular, the engine nozzles' air ejector system identifiable by the characteristic air intake scoops on the rear fuselage was deleted from c/n 1015330 onwards and removed from previously manufactured aircraft in the course of overhauls. From Batch 6 onwards the Su-15TM was equipped with the upgraded SAU-58-2 AFCS enabling automatic interception of low-flying targets (which was beyond the capabilities of the earlier SAU-58). The weapons complement was augmented with two R-60 (AA-8 *Aphid*) 'dog-

Su-15TM '05 Blue' was used in another test programme involving weapons application, as evidenced by the cine camera 'egg' just aft of the radome.

Two excellent views of an operational Su-15TM. Note the wing leading-edge kink in head-on view.

'01 Blue', the U-58TM, was the production prototype of the Su-15UM combat trainer. Here it is seen at Akhtoobinsk in 1976 during state acceptance trials.

A production Su-15UM coded '30 Blue' shows the deployed forward vision periscope in the instructor's cockpit. Note the nose-high ground attitude.

fight missiles' which had passed their trials in late 1974; the R-60s were carried on extra pylons inboard of the existing ones, and the new weapon was integrated *in situ* from 1979 onwards. From c/n 0915311 onwards production Su-15TMs were equipped with new PU-2-8 launch rails which could be easily replaced with BD3-57M bomb racks (**bah**lochnyy der**zhah**tel' – beam-type rack), and appropriate changes were made to the electric system. Finally, at the insistence of the military the Taifoon-M radar was upgraded as part of the measures to offset the damage done by Lt. Viktor I. Belenko's notorious defection to Japan in a MiG-25P on 6th September 1976. The update was a success (see below) but it is not known with certainty if operational Su-15TMs were thus upgraded.

In late 1974 the OKB started design work on a new combat trainer which received the manufacturer's designation **U-58TM**. Developed to meet an Air Force requirement for a trainer version of the Su-15TM, the aircraft was intended for training PVO pilots in flying techniques, aerobatics and combat tactics. In accordance with a joint MAP/MRP ruling agreed on by the Air Force the Sukhoi OKB was to develop a set of project documents for a combat trainer version of the Su-15TM within the shortest possible time, whereupon the Novosibirsk aircraft factory would build a prototype. The project documents were transferred to plant No.153 in the spring of 1975, but it was not until 16th July 1975 that MAP finally issued an order officially initiating development of the U-58TM.

The parties concerned chose not to build a prototype, performing the required trials on the first production aircraft. The trainer was based on the airframe of the late-production Su-15TM; unlike the preceding Su-15UT, there was no fuselage stretch, the overall length being the same as the single-seater's. The twin-wheel nose gear unit was also retained. Remarkably, the provision of a second cockpit did not incur a reduction of the fuel tankage, the space for it being provided solely by deleting some equipment items. The internal fuel capacity was 6,775 litres (1,490.5 Imp gal); additionally, two 600-litre (132 Imp gal) drop tanks could be carried.

In addition to a full set of controls and flight instruments, the rear cockpit featured a special control panel allowing the instructor to deactivate some of the instruments in the trainee's cockpit, simulating hardware failures. The canopy was similar to that of the Su-15UT, the rear section incorporating a retractable forward vision periscope. The area of the horizontal tail was increased slightly to address the elevator authority problem typical of the Su-15.

Equally remarkably, the U-58TM's empty weight of 10,635 kg (23,445 lb) was lower than the single-seater's. This was accomplished by deleting much of the Su-15TM's avionics, namely the radar, the SAU-58-2 AFCS, the Lazoor'-M command link system, the SPO-10 radar warning receiver and the RSBN-5S SHORAN. The avionics included an R-832M radio, an ARK-10 ADF, an RV-5 low-range radio altimeter, an MRP-56P marker beacon receiver, a KSI-5 compass system and an AGD-1 artificial horizon, as well as an SPU-9 intercom and an MS-61 CVR.

The customer required the trainer to retain a measure of combat capability so that weapons training could be performed. Given the lack of radar, this requirement could be met by using heat-seeking missiles; thus the U-58TM featured a weapons control system compatible with the R-98T medium-range AAM and the R-60 short-range AAM; additionally, UPK-23-250 pods could be carried on the fuselage hardpoints.

The 'second-generation' trainer variant received the service designation **Su-15UM** (oo**cheb**nyy, moderni**zee**rovannyy – trainer [version], upgraded). Aptly coded '01 Blue', the production prototype of the Su-15UM (c/n 0115301) was rolled out at Novosibirsk-Yel'tsovka in the spring of 1976; Yuriy K. Kalintsev was appointed engineer in charge of the tests. The maiden flight took place on 23rd April with factory test pilots Vladimir T. Vylomov and V. A. Belyanin at the controls. Five days later the aircraft was flown to Zhukovskiy where OKB test pilots Yevgeniy S. Solov'yov and Yuriy A. Yegorov performed an abbreviated manufacturer's test programme comprising only 13 flights.

On 23rd June the Su-15UM production prototype was turned over to the military for state acceptance trials; these were rather brief, lasting only five months, and were completed on 25th November with good results. At this stage the aircraft was made a total of 72 flights. The State commission's report said that the new two-seater was suitable for training aircrews in take-off and landing techniques, all flight elements and certain elements of combat tactics; the Su-15UM was recommended for production and service.

The Novosibirsk aircraft factory produced the new trainer in 1976-1980 under the product code *izdeliye* 43. The type was cleared for service without the usual Council of Ministers directive – a Ministry of Defence order was all it took. The Su-15UM's NATO reporting name was *Flagon-G*.

The only major change made to the Su-15UM in the course of production was that the RSBN-5S SHORAN was added after all; new-build examples were equipped with the system from batch 4 onwards and deliveries commenced in 1978. The final three trainers off the line were delivered fairly late; two of them (c/ns 0415343 and 0415344) left the plant in February 1981, while Su-15UM c/n 0415345 – the very last *Flagon* built – did not make its first flight until 14th February 1982, a full year after the rollout. The honour of making the 'last first flight' of a Su-15 fell to factory test pilots Igor' Ya. Sushko and Yuriy N. Kharchenko.

The Su-15's overall production run amounted to some 1,300 aircraft.

Experimental Versions and Testbeds

Between 1966 and 1975 a number of Su-15s were used in various research and development programmes – mostly as avionics and equipment testbeds, verifying items which found use on later versions of the *Flagon*. 26 such R&D programmes were completed in 1966-69 alone. For instance, in early 1965, when the T58D-1 had completed a brief flight test programme with the new double-delta wings, the OKB decided to use this aircraft as a propulsion systems testbed and a short take-off and landing (STOL) technology demonstrator in conjunction with the development of the T-58M low-altitude attack aircraft. (The latter designation proved to be short-lived; the T-58M, later redesignated T-6, was a totally unrelated design that evolved into the Su-24 *Fencer* tactical bomber and lies outside the scope of this book.) This involved installing small turbojet engines vertically inside the fuselage to generate lift. The lift-jet concept was quite popular then both in the Soviet Union and in the West. On 6th May 1965 MAP issued an order requiring the Sukhoi OKB to build and test a proof-of-concept vehicle in order to verify the STOL technology using lift-jets.

The project was ready by mid-year; the extensive conversion involved remanufacturing the centre fuselage to accommodate three 2,350-kgp (5,180-lbst) Kolesov RD36-35 turbojets installed in a bay between the cruise engines' inlet ducts with the axes inclined forward 10°. The lift engines breathed through two large dorsal intakes with aft-hinged doors (one for the foremost engine and one for the other two); the exhaust aperture was closed by movable louvres which directed the jet aft to add a measure of forward thrust on take-off or forward to slow the aircraft down on landing. In cruise flight (in 'clean' configuration) the air intakes and exhaust louvres closed flush with the fuselage skin. All fuel was now carried in the wing tanks – and that means less fuel and three more engines guzzling away at it. But then, range and endurance were not crucial for a pure technology demonstrator that was not meant to operate far away from its base.

Designated **T-58VD** (*verti**kahl'**nyye **dvigateli*** – 'vertical engines', i.e., lift-jets), the rebuilt aircraft was completed in late 1965, commencing tethered tests on a purpose-built ground rig at the Sukhoi OKB's premises. The rig featured an 'open-air wind tunnel' – a turboprop engine driving ducted propellers emulated the slipstream at flight speeds up to 400 km/h (248 mph), creating proper operating conditions for the lift engines.

The T-58VD made its first flight at Zhukovskiy on 6th June at the hands of Yevgeniy S. Solov'yov; later it was also flown by Sukhoi OKB chief test pilot Vladimir S. Il'yushin. The lift created by the auxiliary engines reduced the unstick speed from 390 to 285 km/h (from 242 to 177 mph) and the landing speed from 315 to 225 km/h (from 195 to 139 mph); the take-off run was shortened from 1,170 to 500 m (from 3,840 to 1,640 ft) and the landing run from 1,000 to 560 m (from 3,280 to 1,840 ft) – an impressive result. However, the location of the lift engines was not the optimum one, as the thrust of the forward engine caused a strong tendency to pitch up during landing approach; the problem was solved by using only the centre and rear lift-jets for landing.

On 9th July 1967 the T-58VD participated in the airshow at Moscow-Domodedovo, giving a short take-off and landing

The T-58VD STOL technology demonstrator parked at the Sukhoi OKB's flight test facility in Zhukovskiy, still coded '58 Red'. The tandem air intakes for the lift engines can be seen ahead of the fin. Note the double-delta wings.

Here the T-58VD is shown during tethered tests on a special rig at the OKB's premises in Moscow, with an NK-12 turboprop driving a shrouded fan to simulate the slipstream. Note the Cyrillic '58VD' (58ВД) tail titles and the open brake parachute container. The bulge replacing the port upper airbrake probably houses test equipment.

No longer wearing a tactical code, the T-58VD is pictured at a dirt airstrip during trials.

demonstration; after that, the STOL version received the reporting name *Flagon-B*. The results obtained with this aircraft were incorporated into the design of the T6-1 strike aircraft prototype. However, the short-field performance afforded by the latter's lift engines did not justify the disadvantages they incurred (reduced fuel capacity, increased fuel consumption and weight penalty). Hence the T-6 was radically reworked to feature variable-geometry wings which gave the desired results – but that's another story.

While we are on the subject of propulsion testbeds, the fourth production Su-15 *sans suffixe* (c/n 0115304) was used by LII as a **testbed for the Tumanskiy R11F3-300 engine** featuring a special combat rating to improve performance at low altitudes. The **T-58-95** testbed (c/n 0415302) had the starboard R11F2S-300 engine replaced with a flight-cleared prototype **Gavrilov R13-300** afterburning turbojet, the last two digits of the designation referring to the development engine's product code (*izdeliye* 95). Suitably fitted out with data recording equipment, the aircraft was operated by LII where manufacturer's tests of the R13-300 were held in 1967-68. Another Su-15 ('11 Red', c/n 0715311) was fitted with two R13-300 engines by mid-December 1968; After initial flight tests the aircraft was transferred to GK NII VVS in March 1969 The trials showed that the service ceiling, acceleration time, effective

Close-up of the aircraft's centre fuselage with the lift engines' air intakes and exhaust flaps open. Note also the open auxiliary blow-in door of the port engine.

The T-58VD climbs away as it gives a STOL demonstration at the 9th July 1967 airshow at Moscow-Domodedovo.

The T-58VD seen on final approach after its demo flight at Domodedovo. Note the non-standard 'towel rail' aerial under the nose.

The T-58VD streams its brake parachute after landing at Domodedovo.

range, combat radius and field performance had improved thanks to the new engines. Later, two more Su-15s *sans suffixe*, including '37 Red' (c/n 1115337), were re-engined with R13-300s, undergoing tests in 1969-70 and 1970-71 respectively. The latter aircraft featured wider air intakes to cater for the engines' greater mass flow and prevent flame-outs. This intake design was also tested on a further production Su-15 (c/n 1315340) modified in 1970; For security reasons this aircraft was referred to at the Sukhoi OKB as 'aircraft 0009'. The aircraft showed encouraging performance and the new intake design was introduced on the production Su-15TM, together with the R13-300 engines.

On 25th February 1971 the Council of Ministers issued a directive followed by a joint MAP/Air Force ruling. These documents required the Sukhoi OKB to re-engine the Su-15TM with Gavrilov R25-300 afterburning turbojets rated at 4,100 kgp (9,040 lbst) dry and 6,850 kgp (15,100 lbst) reheat, with a contingency rating of 7,100 kgp (15,650 lbst). Development work began in 1972, the aircraft receiving the in-house designation **T-58bis** and the provisional service designation **Su-15bis**. The prototype was converted in Novosibirsk from the fifth production Su-15TM ('25 Blue', c/n 0315306), making its first flight on 3rd July with Vladimir S. Il'yushin at the controls. Manufacturer's flight tests continued until 20th December; the Su-15bis showed a marked improvement in acceleration and top speed at low and medium altitudes over the standard R13-300 powered aircraft, especially with the engines at contingency

Here the T-58VD is seen as an instructional airframe at the Moscow Aviation Institute in the 1970s.

'25 Blue', the sole prototype of the Su-15*bis*, seen during trials.

rating. State acceptance trials took place between 5th June and 10th October 1973 with good results; the contingency rating allowed targets flying at up to 1,000 km/h (620 mph) to be intercepted in pursuit mode. The Su-15*bis* was recommended for production but, ironically, it never achieved production status; conversely, the R25-300 engine did, powering the mass-produced MiG-21*bis* tactical fighter.

A few *Flagons* found use as avionics testbeds. LII used a specially modified Su-15 ('16 Blue', c/n 0615316) as a testbed for passive ECM and infra-red countermeasures (IRCM) equipment, verifying almost all chaff/flare dispenser types used by the

Soviet Air Force, as well as 50-mm (1.96-in) PPR-50 chaff cartridges (*peeropatron rahdiolokatsionnyy*) and PPI-50 magnesium flares (*peeropatron infrakrahsnyy* – infrared flare). Another Su-15 (identity unknown) was used in 1968-69 to test the new R-832M *Evkalipt*-SM (Eucalyptus-SM) communications radio.

In May 1972 the Leningrad-based NII-131 of the Ministry of Electronics Industry (MRP – *Ministerstvo rahdioelektronnoy promyshlennosti*), aka LNPO Leninets (*Leningrahdskoye naoochno-proizvodstvennoye obyedineniye* – 'Leninist' Leningrad Scientific & Production Association), converted the abovementioned Su-15 '11 Red' (c/n 0715311) into an avionics testbed. The nose radome housed a prototype *Rel'yef* (Terrain profile) radar developed for the T-6 (Su-24); this radar enabled automatic terrain-following flight during low-level air defence penetration. The aircraft was designated **T-58R** or **SL-15R**, the SL standing for *samolyot-laboratoriya* (laboratory aircraft) and the R suffix referring to the Rel'yef radar.

In the wake of Lt. Viktor I. Belenko's infamous defection to Japan in a MiG-25P on 6th September 1976, the Soviet Council of Ministers issued a directive in November 1976 requiring measures to be taken in order to minimise the damage done by Belenko's treason. Accordingly the Su-15TM received a modest upgrade of the radar. In early 1977 two production Su-15TMs (c/ns 1315349 and 1415307) were set aside as testbeds under this programme, undergoing tests in June-October that year. It took 12 months to perfect the modified weapons control system, and in 1978 the modifications received approval for incorporation on in-service aircraft. In keeping with the same directive another production Su-15TM was modified for testing the new SRO-1P *Parol'*-2 (Password-2) IFF suite comprising an *izdeliye* 623-1 interrogator and an *izdeliye* 620-20P transponder, which eventually became standard on Soviet military aircraft.

Two specially modified Su-15s were used for extensive aerodynamics research to investigate the aircraft's stability and control characteristics in certain flight modes. LII twice held spinning tests of the Su-15 – in 1968 with the pure-delta version and in 1973 with the double-delta version; Oleg V. Goodkov was project test pilot in both cases. The second aircraft ('37 Red', c/n 1115337) was equipped with PPR-90 spin recovery solid-fuel rockets (*protivoshtopornaya porokhovaya raketa*), enabling high-alpha and spinning tests to be held. A similar spinning test programme was performed by GK NII VVS, with N. V. Kazarian as project test pilot. The tests showed that the aircraft vibrated strongly as it approached critical angles of attack, warning the pilot that he was 'pushing it too far'. Actually the Su-15 could enter a spin only due to a grave piloting error or if the spin was intentional. The spinning characteristics of the two versions were similar, but the double-delta version was more stable during the spin.

In 1968 the Sukhoi OKB started work on modified wings featuring a sharp leading edge; by April 1973 this work materialised in the **T-58K** research vehicle converted from the fourth production Su-15 (the K stood for [*modifitseerovannoye*] *krylo* – [modified] wings). The boundary layer fences and missile pylons were deleted and a new extended leading edge section with a sharper profile was installed; the BU-220 tailplane actuators were replaced by BU-250 units, and part of the standard avionics was replaced by test equipment and ballast. The T-58K underwent trials at LII in 1973-74. Another aerodynamics research vehicle was the 16th production Su-15T (c/n 0215306) which was fitted with increased-area tailplanes at the end of 1972, undergoing a special test programme in 1972-73 with a view to improving the fighter's handling in landing mode with the BLCS activated.

Su-15 '16 Blue' (c/n 1615361) served as a testbed for IRCM equipment. Note the photo calibration markings on the nose and tail.

Service pilots kept complaining that that the Su-15's lateral stability was poor, especially during the landing approach. Hence, when turned over to GK NII VVS for testing in 1974 the second production Su-15TM (c/n 0303) was converted into a control system testbed with a trim mechanism in the aileron control circuit and a lateral stability augmentation system. The new features received a positive appraisal and were introduced on the production line. Later this aircraft was used to test an increased-area horizontal tail, which was also recommended for production – too late, as production had ended by then.

Later, in 1980, the Sukhoi OKB converted Su-15 c/n 1115328 into a **control configured vehicle (CCV)** with variable in-flight stability and controllability parameters. For the first time in Soviet practice the aircraft featured a side-stick controller; the standard control stick was retained and the pilot was able to switch the control system from one stick to the other as required. The CCV underwent tests at LII in 1981-82; unfortunately the aircraft crashed on 11th November 1982 before the test programme could be completed.

Weapons testbeds were one more category. In 1978 Su-15T c/n 0215306 was converted into a weapons testbed to support the development of the Sukhoi T-10 *Flanker-A* fourth-generation fighter (the immediate precursor of the world-famous Su-27). Designated **L.10-10** (i.e., 'flying laboratory' No.10 under the T-10 programme), the aircraft had the standard PU-1-8 missile launch rails replaced with APU-470 launch rails developed for the new K-27E advanced medium-range AAM. The actual test launches of these missiles took place in 1979, the missile eventually entering production as the R-27.

Originally the Su-15's armament consisted solely of two AAMs; yet the military kept requesting that cannon armament be incorporated as well. Initially the Sukhoi OKB intended to use a single GP-9 standardised centreline pod (*gondola **push**echnaya* – gun pod) housing a 23-mm (.90 calibre) Gryazev/Shipoonov GSh-23 twin-barrel cannon with 200 rounds. After a series of tests this was found suitable for the *Flagon*, and the Novosibirsk plant even built ten Batch 12 Su-15s with the appropriate attachment fittings and connectors. (However no GP-9 pods were ever delivered to Su-15 units.) By then, however, the military had changed their requirements; the UPK-23-250 pod housing the same cannon with 250 rounds became the Su-15's standard cannon armament. In addition to the greater ammunition supply, the UPK-23-250 could be fitted and removed extremely easily (unlike the GP-9); most importantly, two such pods could be carried on the Su-15's fuselage hardpoints, giving twice the firepower. A production Su-15 (c/n 1115342) was set aside for cannon armament tests, passing its state acceptance trials in March-September 1971; the GP-9 pod was also fitted to another Su-15 (c/n 0515328). Even though aiming accuracy with the Su-15's standard K-10T sight left a lot to be desired, the installation was recommended for use against both air and ground targets. Since production of the basic Su-15 *sans suffixe* had ended by then, these aircraft were retrofitted with UPK-23-250 pods in service.

The OKB did not give up on the idea of fitting an internal cannon to the Su-15; now a GSh-23L cannon was to be mounted on the fuselage centreline aft of the nosewheel well, as on the Mikoyan MiG-23 *Flogger* fighter. The third production Su-15TM (c/n 0315304) was modified for testing this cannon installation, eventually undergoing state acceptance trials in 1973. The results were very similar to those obtained with the standard UPK-23-250 pods; thus the built-in cannon was recommended for production. Still, gunnery accuracy was rather poor because the standard K-10T sight was ill-suited for working with cannons, since there was no room for a specialised gunsight, the built-in cannon never found its way to the production line.

When the Su-15TM armed with R-98M missiles entered service it was decided to adapt existing Su-15s *sans suffixe* with the Oryol-D58 radar for using this missile, which was originally designed for use with the Taïfoon radar. To this end a late-production Su-15 (c/n 1415301) was converted in 1975, successfully passing a special test programme. In the early 1970s the **Mol**niya (Lightning) OKB brought out the R-60 agile short-range IR-homing AAM, which was selected as the main close-in weapon for all Soviet fighter types. After the R-60 had been integrated on the Su-15TM it was decided to up-arm the Su-15 *sans suffixe* with this missile as well. A suitably modified early-production Su-15 (c/n 0615327) served for testing this upgrade successfully in 1978-

The experimental installation of a GP-9 cannon pod on a Su-15 (c/n 0515328).

79, the Air Force's aircraft repair plants started upgrading Su-15s with inboard pylons for two R-60s. A configuration with four R-60s on APU-60-2 paired launchers (*aviatsionnaya pooskovaya oostanovka* – aircraft-mounted launcher) was also tested but did not find its way into service.

The incorporation of bomb armament on the Su-15 for use as a tactical strike aircraft had been repeatedly delayed ever since the beginning of the state acceptance trials. Finally, by August 1974 a production Su-15TM (c/n 1007) was converted to feature four BD3-57M bomb racks instead of the standard wing and fuselage pylons, with appropriate changes to the electrics. The K-10T sight was equipped with a tilting mechanism enabling attacks against ground targets. The aircraft was tested in two stages; Stage A (October 1974 – May 1975) was to verify the bomb armament, while Stage B (June-December 1975) was concerned with the cannon and rocket armament. The trials were successful, the report stating that the armament fit rendered the aircraft suitable for use against pinpoint ground targets. As a result, late-production Su-15TMs were equipped with PU-2-8 launch rails which could be easily exchanged for BD3-57M racks.

Until the 1970s the Soviet Air Force required virtually all Soviet fighter types to be capable of carrying tactical nuclear bombs – originally because no supersonic tactical bombers existed and then probably as an insurance policy in case the bombers were destroyed by enemy strikes. The Su-15 also 'fell victim' to this trend; a single *Flagon* was modified as the **T-58N** tactical nuclear strike aircraft (*nositel' spetsizdeliya* – lit. 'carrier', i.e., delivery vehicle for *special stores*, as the nuclear bombs were coyly referred to). No details have been disclosed to date.

In the early 1960s, when the Cuban missile crisis put the world on the brink of nuclear war, the Soviet Armed Forces paid much attention to dispersing troops in order to make them less vulnerable to enemy strikes. In the case of military aviation this meant operations from unpaved tactical and reserve airstrips – for which the VVS and the PVO were totally unprepared. This led the military to demand insistently that all tactical aircraft types should be capable of operating from semi-prepared dirt strips. Accordingly, building on previous experience with experimental versions of the Su-7B, the Sukhoi OKB developed a special landing gear for the Su-15 with a view to exploring the possibility of operating the interceptor from such strips. The designers opted for a mixed arrangement with skids on the main gear units and a wheeled nose unit. The second prototype *Flagon-A* (the T58D-2) was converted in the first half of 1965 and redesignated **T-58L**, the suffix denoting *lyzhnoye shassee* – skid landing gear. The new main gear units (with appropriately modified wheel wells and gear doors) could

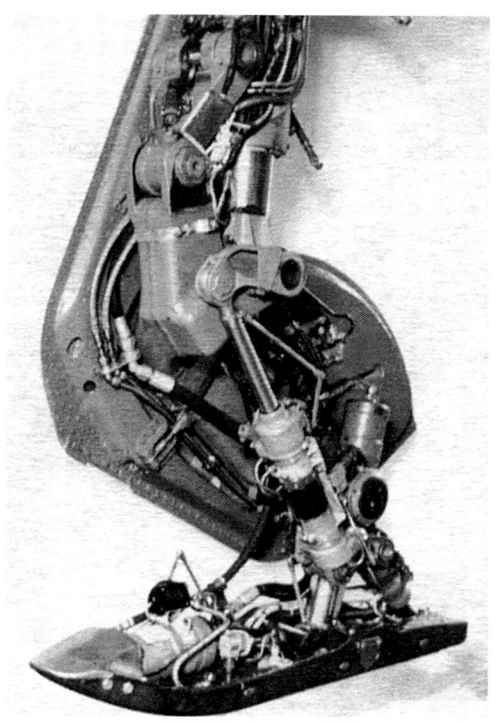

The starboard main gear unit of the T-58L with the skid fitted.

be quickly reconfigured from wheels to skids and back again; the skids featured a lubrication system to facilitate movement on grass and packed earth. The standard castoring nosewheel was replaced by a steerable nose gear unit. Provisions were made for installing jet-assisted take-off (JATO) solid-fuel rocket boosters, and a new KS-4 ejection seat was fitted. The T-58L first flew on 6th September 1965 at the hands of Vladimir S. Il'yushin. Until the mid-1970s the aircraft underwent extensive testing on various semi-prepared grass, dirt and snow strips in various climatic zones with the participation of GK NII VVS; it was flown by OKB test pilots and GK NII VVS pilots. In the course of the tests the T-58L was refitted with a new, taller nose gear unit featuring twin KN-9 non-braking wheels; the purpose was to increase the angle of attack on take-off (thereby increasing lift and shortening the take-off run), raise the air intakes higher above the ground (thereby reducing the risk of foreign object ingestion) and improve ground manoeuvrability. The skid landing gear was not introduced on the production model, as the vibrations experienced on uneven runways subjected the avionics and armament to high G loads which could ruin them; also, the missiles were liberally spattered with dirt, which likewise could put them out of action. As for the T-58L, it became an instructional airframe and subsequently an exhibit at the Air Force Museum in Monino.

In-flight refuelling (IFR) takes a special place among the research and development work performed by the Su-15. The Sukhoi OKB's first involvement with IFR dates back to 1971, when the need arose to increase the range and combat radius of the Su-24 tactical bomber which was undergoing trials at the time. By the end of the year the OKB pre-

The T-58L development aircraft was converted from the second prototype Su-15 *sans suffixe* (T58D-2), hence the code '32 Red'. The twin-wheel nose gear unit is evident, as is the original area-ruled fuselage. Note the Cyrillic '58-L' (58-Л) tail titles.

pared a set of project documents envisaging the installation of an IFR system on what was then the T-58M. Unlike the Sukhoi OKB, LII and the *Zvezda* (Star) OKB headed by Guy I. Severin had accumulated a wealth of experience in developing, testing and using various IFR systems. The Zvezda OKB had been developing such systems since the late 1960s. In 1971-73 its specialists designed and tested the principal components of a probe-and-drogue refuelling system; the programme was codenamed *Sakhalin* after an island in the Soviet Far East – perhaps the implication was that 'with this system our aircraft will be able to reach Sakhalin non-stop'!

In keeping with the recommendations of MAP's Scientific & Technical Council the Sukhoi OKB decided to hold a series of tests in advance so that the probe-and-drogue system would be fully mastered by the time the bomber was ready to take it. In October 1973 the VVS issued a specification in which three Sukhoi types – the Su-15TM, the Su-17M/Su-17M2 and the Su-24 – were stated as possible tanker aircraft fitted with 'buddy-buddy' refuelling pods. The pod, or hose drum unit (HDU), was designated UPAZ-1 (*oonifitseerovannyy podvesnoy agregaht zaprahv*ki – 'standardised suspended (i.e., external) refuelling unit'). It had a 26-m (85 ft 3 in) hose and a flexible 'basket'

drogue. The hose drum was powered by a ram air turbine (RAT) with an intake scoop on the port side which was normally closed. A second air intake in the nose closed by a movable cone was for an RAT driving a generator for the electric transfer pump. Normal delivery rate was 1,000 litres (220 Imp gal) per minute, increasing to 2,200 litres (484 Imp gal) in case of need.

Pursuant to a joint MAP/Air Force ruling the Sukhoi OKB allocated two Su-15s – the first pre-production aircraft ('01 Red', c/n 0015301) and an uncoded early production aircraft (c/n 0215306) – as **IFR system testbeds**, one aircraft acting as the tanker and the other as the receiver. Both aircraft had by then a long history as 'dogships', being used for testing various systems and equipment. Additionally, LII allocated one of its Su-15s ('37 Red', c/n 1115337) for testing the Sakhalin-1A system and evolving IFR techniques.

By December 1972 Su-15 '01 Red' had been fitted with a dummy UPAZ-1A HDU for aerodynamic testing (the pod featured a functional hose drum allowing the drogue to be deployed and retracted). On 19th December OKB test pilot Vladimir A. Krechetov performed the first test flight, assessing the aircraft's stability and handling with this rather bulky external store.

In 1973 the OKB issued a set of documents for the adaptation of the standard Su-15's fuel system to accept the UPAZ-1A pod. This involved installation of additional fuel pumps in the Nos. 1, 2 and 3 fuselage tanks, modifications to the electric system and replacement of the standard TRV1-1A fuel metering kit with a new TRK1-1 kit. The radar display was removed to make room for the HDU's control panel. The conversion was performed in April-July 1974 by the OKB's Novosibirsk branch. On 4th July 1974 Sukhoi OKB test pilot Aleksandr S. Komarov made the first check flight from Novosibirsk-Yel'tsovka, whereupon the modified aircraft was flown back to Zhukovskiy for testing.

The second 'tanker' aircraft was Su-15 '37 Red' built in February 1970 and delivered new to LII. It had double-delta wings and an operational BLCS; also, it could be fitted with R11F2SU-300 or R13-300 engines. In late 1973 the fighter was modified in house by LII for installation of an operational UPAZ-1A HDU. This aircraft served for testing the pod proper and for verifying the approach and contact technique. Contact would be made in 'dry' mode (without actual fuel transfer), hence the scope of the modification work on this aircraft was much smaller, involving removal of the radar, which was replaced by ballast, and installation of the HDU control panel, the 'wet' centreline pylon and appropriate data recording equipment. In 1974 LII test pilots conducted a series of tests on this aircraft, checking the operation of the HDU (ie, hose deployment/retraction and the operation of the pod's other systems).

Su-15T c/n 0215306 was converted into the receiver aircraft under the Sakhalin programme at the Sukhoi OKB in late 1973. Again the conversion was quite extensive, as the correct fuel transfer sequence had to be ensured in order to keep the CG within the allowed limits. The aircraft was equipped with a TRK1-1 fuel metering kit and a fixed L-shaped IFR probe offset to starboard ahead of the cockpit. Cine cameras were fitted for filming the contact with the drogue; a flashing beacon was installed on the upper centre fuselage for synchronising the operation of all photo and cine cameras capturing the refuelling sequence. Since the refuelling operation imposed considerable stress on the pilot, the receiver aircraft was fitted with special *Koovshinka* (Water lily) medical equipment recording the pilot's life signs (pulse and breath) and recording the distribution of his concentration during various stages of the process. Lastly, the aircraft was refitted with a twin-wheel nose gear unit.

Stage One of the tests (performed jointly by the Sukhoi OKB and LII) was meant to develop the optimum approach and contact technique. Before the contacts with the tanker could begin, the probe-equipped Su-15T made a series of test flights to see how the IFR probe affected stability and handling. The actual flight test programme commenced on 31st May 1974 when the first two flights were made. The rendezvous with the tanker and attempted contacts were made at 8,200 m (26,900 ft) on the first occasion and about 6,000 m (19,680 ft) on the second occasion; the speed was 550 km/h (340 mph) in both cases. Both attempts ended in failure; moreover, on the second try the fuel transfer hose was torn as the receiver aircraft manoeuvred. A pause ensued while the engineers made corrections to the piloting technique during the final approach phase, the flights resuming only on 24th December 1974. During this period the Zvezda OKB revised the HDU, increasing the length of the hose to 27.5 m (90 ft 2 in). In the meantime project test pilot Yevgeniy S. Solov'yov trained on a purpose-built simulator and in the actual aircraft (the approach to the tanker was simulated on the ground). The effect of the additional training was felt immediately; as early as 14th January the modified Su-15T made the first stable contact with the tanker. This flight marked the end of the phase involving LII's 'buddy tanker' (Su-15 c/n 1115337), as the OKB's own 'buddy tanker' (Su-15 c/n 0015301) had been completed by then.

Solov'yov remained the receiver aircraft's pilot at Stage Two; the tanker was flown by Aleksandr S. Komarov and Vladimir A. Krechetov. The beginning of the programme's main phase was delayed until 21st January 1975. The first contacts were made in 'dry' mode at altitudes of 2,000-7,500 m (6,560-24,600 ft) and speeds of 480-660 km/h (300-410 mph); each mission involved two to seven attempts. This stage continued until the end of February, after which another lengthy hiatus followed due

to the need to analyse the results obtained and prepare for the next phase. The flights resumed in June 1975 in 'dry' mode; in early July the HDU's drogue was modified, allowing fuel to be transferred. An immediate problem arose: the drogue turned out to be defective, causing fuel leaks after contact had been made; a new, properly made drogue took care of the problem. The first successful transfer of 250 kg (550 lb) of fuel took place on 30th July 1975; the following day two more fuel top-ups were made.

By early December 1975 the Sukhoi OKB and the Zvezda design bureau had ironed out all the bugs discovered in the course of the tests. After that, another four flights involving three fuel top-ups were made between 10th and 23rd December; thus the test programme was successfully completed and the Sakhalin-1A IFR system was recommended for service. Shortly afterwards Su-15 c/n 0015301 was returned to the PVO as time-expired, becoming a ground instructional airframe at a PVO Junior Aviation Specialists School in Solntsevo just outside Moscow. After the closure of the school in 1991 the aircraft was part of the now-defunct open-air aviation museum at Moscow-Khodynka. As for Su-15 c/n 1115337, it continued in use with LII, participating in other research programmes until it crashed fatally near Lookhovitsy, Moscow Region, on 24th December 1976; the cause was never determined.

Opposite page:
Seen here in front of the Sukhoi OKB hangar at Zhukovskiy, Su-15 '01 Red' (c/n 0015301) was the first of two to be fitted with an UPAZ-1A 'Sakhalin' hose drum unit during IFR system trials.

The other Su-15 equipped with the UPAZ-1A HDU was '37 Red' (c/n 1115337); note the double-delta wings.

The receiver aircraft (c/n 0215306) in front of the Sukhoi OKB hangar at Zhukovskiy. Note the twin-wheel nose gear unit retrofitted during the IFR system test programme and the photo calibration markings on the nose.

Another view of the same aircraft, showing clearly the fixed offset IFR probe and the cine camera fairing on the fin. Note also the late-model air data boom as fitted to the Su-15TM.

This page:
A still from the footage filmed by the receiver aircraft's fin tip camera as Su-15 c/n 0215306 makes contact with the 'buddy' tanker.

Here the two aircraft are seen from a chase plane, with Su-15 '01 Red' acting as the tanker.

Unbuilt Projects

During 1968-69 the Sukhoi OKB considered the idea of equipping the **Su-15 with a built-in cannon** in the starboard wing root. Provisionally designated *izdeliye* 225P, the experimental fast-firing 23-mm Gatling cannon was developed by the Fine Machinery Design Bureau (Kon**strook**torskoye *byuro* **toch**novo ma**shin**ostro**yen**iya) in Tula. Yet accommodating the ammunition box inside the Su-15's thin wing proved to be an insurmountable task and the idea was dropped.

In February 1966 the Sukhoi OKB considered re-engining the Su-15 with Solov'yov D-30 turbofans. Developed by OKB-19 in Perm' for the Tupolev Tu-134 *Crusty* short-haul airliner, this commercial engine delivered 6,800 kgp (14,990 lbst), although an even more powerful afterburning version was probably envisaged for the fighter. However, installing D-30s would require major structural changes, which would disrupt production, and the project – provisionally designated **T58D-30** to indicate the engine type – was shelved.

In the summer of 1969 the Sukhoi OKB looked into the possibility of transforming the T-58 into a fully-fledged attack aircraft. The SPB ground attack aircraft project (*samolyot polya boya* – battlefield aircraft), the precursor of the famous Su-25 *Frogfoot*, unexpectedly ran into an in-house competitor: Anatoliy M. Polyakov proposed an aircraft designated **T-58Sh** (*shtoormovik* – attack aircraft). Billed as an 'in-depth upgrade' of the T-58 interceptor, this was for all intents and purposes a new aircraft. The forward fuselage was drooped to provide adequate downward visibility; the wings were also new, featuring a trapezoidal planform with reduced leading-edge sweep and greater area. The cockpit section and the engines were protected by armour and the fuel tanks were self-sealing for higher survivability. The radar and other avionics associated with the interceptor role were replaced by an ASP-PF gunsight, a PBK-2 bomb sight optimised for lob-bombing (*pritsel dlya bombometaniya s kabreerovaniya*) and a Fon (Background) laser ranger. The T-58Sh featured eight weapons hardpoints and a built-in *izdeliye* 225P cannon. The external stores options included bombs of up to 500 kg (550 lb) calibre, unguided rockets, Kh-23 laser-guided air-to-surface missiles, UPK-23-250 cannon pods and SPPU-17 pods with movable cannons for strafing ground targets in level flight. For self-defence the aircraft would be armed with K-55 and K-60 AAMs. At a 17,500-kg (38,580-lb) take-off weight the aircraft was to lug a 4,000-kg (8,820-lb) ordnance load. However, the T-58Sh did not progress beyond the preliminary design stage, losing out to the more promising T-8 – the future Su-25.

In 1972-73 the Sukhoi OKB proposed an in-depth upgrade of the Su-15, striving to enhance the interceptor's performance by radically improving the aerodynamics. Since the OKB placed high hopes on the ogival wings developed for the T-10 *Flanker-A* fighter, the intention was to use them on the Su-15 as well. The rewinged interceptor bore the in-house designation **T-58PS**; later it was referred to in the Sukhoi OKB's correspondence with MAP and the Air Force as the **Su-19**. A series of wind tunnel tests was held at TsAGI, followed by more detailed research into layouts utilising ogival wings. Estimates showed that the interceptor's performance and agility would be enhanced dramatically; also, the new wings provided room for two additional hardpoints, allowing more short-range AAMs to be carried – a real asset in a dogfight, in which the more manoeuvrable fighter could now engage.

The next step towards improving performance was to equip the prospective Su-19 (T-58PS) interceptor with advanced R67-300 engines instead of R13-300s. The Sukhoi OKB prepared a technical proposal, sending it to MAP and the Air Force for appraisal. According to this document the basic Su-19 could enter flight test in the fourth quarter of 1973, the re-engined version designated **Su-19M** following in the first quarter of 1975. Yet the military showed a complete lack of interest. The OKB then proposed fitting the existing Su-15TM with a new *Poorga* (Snowstorm) fire control radar to give it the requested 'look-down/shoot-down' capability. In addition, the aircraft (provisionally designated **Su-15M**) would be powered by Lyul'ka AL-21F-3 afterburning turbojets rated at 7,800 kgp (17,195 lbst) dry and 11,215 kgp (24,725 lbst) reheat and armed with K-25 AAMs. The PVO top command supported the idea; yet MAP regarded this project with a jaundiced eye and all further work on upgrading the Su-15 had to be abandoned.

Two views of the T-58Sh attack aircraft from the project documents, showing the new swept wings of trapezoidal planform and the drooped 'duck bill' nose.

A drawing of the projected Su-19M interceptor with ogival wings; note the six wing pylons.

The Su-15 in Action

A typical Soviet-era publicity shot of a 'valiant PVO fighter pilot ever ready to take on any intruder' as he poses with his early-production Su-15 *sans suffixe* with pure delta wings.

On the morning of 9th July 1967 huge crowds of Muscovites and visitors to the capital rushed to Moscow's Domodedovo airport where a spectacular airshow – the first in the last six years – was to take place. After the hardships of the Khrushchov era the Soviet aircraft industry and the Air Force were eager to show the nation's new leaders and the public at large that, in spite of the battering they had taken due to Khrushchov's 'missile itch', they were still very much alive and had considerable potential. Hence it was decided to display almost everything the industry had to offer, including aircraft that were still undergoing trials at the time or were just about to enter production and/or service. The latter included the Sukhoi OKB's latest product, the Su-15 interceptor, which received the honour of opening the flying display of the show. Right on schedule a group of five Su-15s flown by the PVO's 148th TsBP i PLS (*Tsentr boyevoy podgotovki i pereoochivaniya lyotnovo sostava* – Combat Training & Aircrew Conversion Centre) pilots and led by Col. P. P. Fedoseyev made a high-speed pass over the improvised grandstand and pulled into a spectacular formation climb, fanning out at the top. The display also included a short take-off and landing demonstration by the T-58VD STOL technology demonstrator flown by OKB test pilot Yevgeniy S. Solov'yov. A while later, Sukhoi OKB chief test pilot Vladimir S. Il'yushin made a flypast in a production Su-15 coded '47 Red' and painted black overall for sheer effect.

From the number of Su-15s participating in the show Western military analysts concluded that the Soviet Union had fielded a new interceptor, and the Su-15 was allocated the NATO reporting name *Flagon*. Western aviation experts made a fairly accurate guess as to the fighter's performance and correctly speculated that the aircraft was powered by Tumanskiy R11F-300 afterburning turbojets. The advent of the Su-15TM with its ogival nose, however, confused the Western experts somewhat; for some reason they decided that the aircraft was fitted with a new radar (which was not far from the truth) and powered by Lyul'ka AL-21F engines (which was absolutely wrong). Codenamed *Flagon-F*, the upgraded aircraft was referred to in the Western press as the 'Su-21'; it was quite some time before the correct designation became known.

As already mentioned in the first chapter, the Su-15-98 aerial intercept weapons system was formally included into the Soviet Air Defence Force inventory in April 1965 upon completion of the state acceptance trials. The system was capable of intercepting targets flying at speeds of 500-3,000 km/h (310-1,860 mph) and altitudes of 500-23,000 m (1,640-75,460 ft). The interceptor was guided towards the target by the Vozdukh-1 automated GCI system until the target came within range of its fire control radar.

The top brass of the Soviet Ministry of Defence and the PVO had a lot riding on the new interceptor which was to replace several obsolete aircraft types in the PVO inventory. Following the usual practice, the 148th TsBP i PLS at Savasleyka AB was the first unit to master the new type; the Centre's 594th UIAP (oo***cheb***nyy istre***bit***el'nyy ***avi***a***polk*** – fighter training regiment) started conversion training for the Su-15 in early 1966. Su-15 production in Novosibirsk was getting under way slowly, and the practice part of the training course could not begin until Sukhoi OKB test pilot Vladimir S. Il'yushin had ferried the second pre-production aircraft (c/n 0015302) from Zhukovskiy to Savasteyka AB on 28th October 1966. By the end of the year the Centre's pilots had made 14 flights in this aircraft.

Deliveries of truly production Su-15s to the 594th UIAP began in January 1967. Since no dual-control trainer version of the Su-15 existed yet, conversion training had to be undertaken using single-engined Su-9U trainers. Even so, the Su-9U was in short supply and was badly needed by the operational units flying the type, and the decision was taken to use swept-wing Su-7U trainers instead. In May 1967 service pilots started taking their conversion training in the Su-7U before making their first solo flights in the Su-15.

In the spring of 1967 a display team was formed at the 148th TsBP i PLS specially for the abovementioned airshow at Moscow-Domodedovo which was to take place in July; the team included 594th UIAP pilots, pilots from the Centre's Command & Control Squadron and a few service pilots. To rehearse the display the team temporarily relocated to Zhukovskiy together with its aircraft. Training flights were made with dummy R-98 missiles and did not go without incident; on 4th July, five days before the show, one of the fighters (c/n 0315304) lost a missile *together with the pylon* after pulling into a step climb. Investigation of the incident showed that the aircraft had exceeded its operational G load limit by far, and it was no wonder that the pylon had broken off. Detailed examination of the airframe revealed substantial permanent structural deformation and the aircraft was declared a write-off; thus the whole affair was actually a non-fatal accident. To prevent further incidents the actual display flight at the show was performed without missiles and with smoke generator pods on the fuselage hardpoints instead of drop tanks.

Deliveries to operational PVO units began in the spring of 1967. The 611th IAP (istre***bit***el'nyy ***avia***polk – fighter regiment) of the Moscow PVO District based at Dorokhovo AB (Yaroslavl' Region) was the first to re-equip. The 62nd IAP based at Bel'bek AB on the Crimea Peninsula, the Ukraine, followed in July; the 54th GvIAP (G***var***dey***skiy istre***bit***el'nyy ***avia***polk** – Guards fighter regiment) at Vainode AB, Estonia, re-equipped shortly afterwards.

Under the conversion training programme one pilot for each Su-15 would be trained within a six-month period to perform combat duty in visual and instrument meteorological conditions in the daytime and in VMC at night. This proved difficult to accomplish; to hasten the training process, qualified flying instructors (QFIs) were seconded to operational PVO units from the 148th TsBP i PLS.

The strategic bombers in service with the US Air Force (primarily the Boeing B-52 Stratofortress) and the Royal Air Force (the V-Bombers), as well as the Hound Dog (USAF) and Blue Steel (RAF) supersonic air-to-surface missiles carried by these aircraft, were envisaged as the principal targets which the Su-15 would have to deal with. As a dogfighting machine the Su-15 was no good, of course, as it lacked the agility – but then, it was not designed with dogfights in mind. The addition of R-60 heat-seeking short-range AAMs did not increase the interceptor's chances in the event of an encounter with enemy fighters but still improved the chances of a 'kill' against a typical target.

By June 1968, 130 Su-15s were in service with eight fighter regiments. A total of 149 pilots were qualified to fly the type but less than 50% of them were fully trained to perform combat duty in IMC in the daytime and only two (!) were able to fly night sorties. During training special attention was given to engagements in head-on mode, as this type of attack was new for Soviet interceptor pilots. Live weapons training at a weapons range near Krasnovodsk involving missile launches commenced in April 1967; the 611th IAP's pilots were the first to do so, firing 47 missiles at various practice targets.

The same 611th IAP was selected to hold the obligatory service trials (evaluation) of the new interceptor, receiving ten Su-15s (mostly Batch 3 aircraft). The trials proceeded from 29th September 1967 to 15th May 1969. During this period the ten fighters made a total of 1,822 flights, including 418 under the actual evaluation programme; two live weapons training sessions were held with the expenditure of 58 AAMs. The service tests basically corroborated the results of the state acceptance trials. However, a number of serious shortcomings was discovered; among other things, the service ceiling fell short of the specifications due to a decrease in engine thrust in the course of the engines' service life, and the interception range was also less than expected.

An early-production Su-15 *sans suffixe* loaded with black-striped inert R-98 missiles is readied for a practice sortie. The packed brake parachute has been loaded but the container doors are still open.

Su-15 '50 Blue' is depicted as it inspects a western military aircraft over international waters close to the Soviet border. Such encounters were common in Cold War days. Note the live R-98T (port) and R-98R (starboard) missiles.

Su-15s often carried a pair of UPK-23-250 cannon pods for close-in engagements, as illustrated by this late-production *Flagon-A* returning from a sortie.

This Su-15 *sans suffixe* (apparently with early-model pure delta wings) has been upgraded to feature two additional pylons for R-60M short-range AAMs, Su-15TM style.

This page:
A late-production Su-15 *sans suffixe* with double-delta wings, likewise retrofitted with two additional pylons, undergoes maintenance outside its hardened aircraft shelter (HAS) at a PVO airbase. The forward avionics bay covers have been removed and an APA-35-2MU ground power unit on a ZiL-130 chassis is connected as the technicians check the fighter's radar.

An upgraded Su-15 *sans suffixe* with pure delta wings (the straight leading edge is clearly visible) armed with R-98R/R-98T missiles and UPK-23-250 cannon pods.

Opposite page:

Top and centre:
A Su-15UT coded '95 Blue' is seen at a Polish airbase during the Su-15's first foreign deployment. A single-seat Su-15TM comes in to land.

Bottom: A pair of Su-15TMs prepares for a night sortie; the pilots are wearing pressure suits and full-face pressure helmets. The base also hosted a MiG-25 unit.

By the end of 1975 the PVO intended to re-equip 41 fighter regiments with the Su-15, whereupon the new interceptor would make up nearly 50% of the PVO's aircraft fleet. However, when Su-15 production ended in early 1976, the type was in service with only 29 units, 18 of them operating the Su-15 *sans suffixe* and 11 units flying the Su-15TM.

To give credit where credit is due, of all fighter types operated by the PVO the Su-15 probably had the highest percentage of successful real-life intercepts of aircraft intruding into Soviet airspace. Its baptism of fire came on 11th September 1970... well, actually the expression 'baptism of fire' is not really applicable because no shots were fired on this occasion. At 0336 hrs Moscow time PVO radar pickets near Sevastopol', the Ukraine, detected a lone aircraft heading north towards the Soviet border at 3,000 m (9,840 ft), and a 'Red Alert' was called. The target was then 260 km (160 miles) southwest of the city; when it approached within 100 km (62 miles) of the border, a 62nd IAP Su-15 scrambled from Bel'bek AB to prevent an incursion. The target turned out to be an elderly Douglas C-47 belonging to the Hellenic Air Force, and when it eventually crossed the border the fighter lined up alongside and rocked its wings in the internationally recognised 'follow me' signal. The Dakota complied, landing at Bel'bek AB. It turned out that the pilot, Lt. M. Maniatakis, had stolen the aircraft from Kania AB on the island of Crete and fled from his homeland where the fascist junta of the Black Colonels had seized power. Maniatakis requested political asylum in the USSR, which was in all probability granted.

Throughout the 1970s the southern borders of the Soviet Union perpetually received the attentions of hostile aircraft coming from Turkey and Iran. The events described below are but a few of the incursions that took place there.

On 7th September 1972 a flight of Turkish Air Force (THK – *Türk Hava Kuvvetleri*) North American F-100 Super Sabres entered Soviet airspace near Leninakan, Armenia (the city is now called Gyumri). Despite flying at ultra-low altitude, the intruders were detected by air defence radars in a timely fashion. Another ploy of the 'bad guys' worked, however – the fighters flew in close formation, appearing on the radarscopes as one heavy aircraft (the USAF had used this tactic in Vietnam); hence only a single 166th IAP Su-15 was scrambled from Sandar AB in neighbouring Georgia to intercept 'it'. The GCI command post operators did not realise that the target was not an 'it' but a 'they' until the Turkish fighters swept over the place with a roar.

The lone Su-15 proved incapable of intercepting its quarry because its radar lacked 'look-down/shoot-down' capability. As a result, the F-100s passed over Leninakan and were fired upon by a heavy machine gun providing anti-aircraft protection for the PVO's radar site but got away unscathed.

On 23rd May 1974 another THK F-100 intruded into Soviet airspace over the Caucasus region with impunity. A Su-15 standing on QRA duty scrambled from the airbase in Kyurdamir, Azerbaijan, but was not directed towards the target because the latter had unwisely intruded in an area defended by an SAM regiment. A missile was fired at the F-100 but missed due to a malfunction in the guidance system.

Eventually, however, the Turks fell victim to the rule 'pride goeth before the fall'. On 24th August 1976 Soviet AD radars detected a target moving in Turkish airspace towards the Soviet border. This was soon identified as a pair of F-100s flying in close formation. No fewer than three Su-15s scrambled this time (two from Kyurdamir and one from Sandar AB), but again they did not manage to get a piece of the action. The fighters had again rashly flown right into a nest of SAMs; this time the PVO crews on the ground did their job well and one of the Super Sabres was shot down. Unfortunately the wreckage fell on the wrong side of the border and the pilot, who ejected, also landed in Turkish territory; the following day the Turks raised hell, accusing the Soviet Union of the 'wanton destruction of a Turkish fighter'.

A while earlier, on 2nd April 1976, a 777th IAP Su-15 flown by Lt (SG) P. S. Strizhak scrambled from Sokol AB on Sakhalin Island to intercept a USAF Boeing RC-135 reconnaissance aircraft which had entered the 100-km territorial waters strip. Shortly after take-off the pilot was redirected towards a new target – a Japanese Maritime Self-Defence Force (JMSDF) Lockheed-Kawasaki P2V Neptune reconnaissance aircraft flying over the Sea of Japan at 2,000 m (6,560 ft) off the southern tip of Sakhalin. Approaching within 5-6 km (3.1-3.7 miles) of the target, the interceptor followed it, flying a parallel course. Apparently Strizhak flipped the wrong switch and inadvertently

fired an R-98R missile at the Neptune, though no order to attack had been given. Realising what he had done, the pilot made a turn just in time, causing the missile to lose target lock-on; the missile passed off the spyplane's starboard wing and self-destructed harmlessly.

One more publicity shot of a pilot gazing skywards as he sits in the cockpit of Su-15TM '81 Blue'. The next aircraft on the flight line, '61 Blue', is being worked on, with a technician in the cockpit and a UGZS-M.AR nitrogen charging vehicle connected.

A pilot wearing an ordinary ZSh-5 'bone dome' helmet with tinted visor climbs into his Su-15TM '55 Blue' at the quick reaction alert (QRA) duty hardstand. The aircraft carries two R-98 missiles and two cannon pods.

At 1457 hrs Moscow time on 23rd December 1979 a Cessna 185 Skywagon light aircraft entered Soviet airspace 175 km (108 miles) south-west of Maryy, Turkmenia (pronounced like the French name Marie), coming from Iranian territory and flying at about 3,000 m. The aircraft was detected by PVO radars three minutes before it crossed the border, and a 156th IAP Su-15 took off almost immediately from Maryy-2 AB to intercept it. The pilot was directed towards the target by GCI stations but failed to spot it because the Cessna was camouflaged (so much for allegations about 'navigation errors'). The radars lost track of the target shortly afterwards and, after circling for a few minutes, the fighter pilot had no choice but to head for home. (Three more Su-15s and a MiG-23M had also scrambled by then, but they were not directed towards the target.) Nevertheless, his mission was accomplished; when (unbeknownst to the Soviet pilot) the interceptor passed directly above the Cessna, its pilots aborted their plan, losing altitude and opting for an emergency landing for fear of being shot down (hence the disappearance of the target from the radarscopes). Eventually they landed on a highway 195 km (121 miles) west of Maryy and were soon arrested by Soviet border troops.

Another incident on the Iranian border occurred on 18th July 1981. An unidentified aircraft flying at about 8,000 m (26,250 ft) briefly entered Soviet airspace but then left it and the pair of 166th IAP Su-15s which had taken off to intercept it was ordered back to base. A few hours later, however, another unidentified aircraft intruded into Soviet airspace; this time a single 166th IAP Su-15 flown by Capt. Valentin A. Kulyapin and armed with two R-98s and two R-60s scrambled to intercept it. The intruder turned out to be a Canadair CL-44D4-6 freighter, LV-JTN (c/n 34), chartered from the Argentinean airline Transporte Aéreo Rioplatense for smuggling weapons – officially 'pharmaceuticals' – from Israel to Iran and flown by a Swiss crew. The fighter pilot gave the customary 'follow me' signals, trying to force it down at a Soviet airfield; instead, the big turboprop started manoeuvring dangerously, making sharp turns in the direction of the Su-15. The pursuit continued for more than ten minutes; eventually Kulyapin received orders to destroy the intruder. Since the border was very close and the target could get away before the fighter could move away to a safe distance for missile launch, Kulyapin chose to ram the target. The attack was skilfully executed; moving into line astern formation, the Su-15 pitched up into a climb, slicing off the CL-44's starboard tailplane with its fin and fuselage. The freighter plummeted to the ground, killing all four occupants; however, Kulyapin's aircraft was

seriously damaged by the collision and the pilot ejected, landing safely not far from the wreckage of both aircraft. This time the wreckage fell on Soviet territory, furnishing irrefutable evidence of a border violation. For this performance Capt. Kulyapin was awarded the Order of the Red Banner.

Gradually, together with the MiG-25P heavy interceptor capable of Mach 2 flight, the Su-15 supplanted the outdated Su-9, Su-11, Yak-28P and MiG-21PFM from the PVO inventory. Su-15s saw service with units stationed in almost all borderside regions of the Soviet Union, the High North and the Far East receiving the highest priority. The Su-15TM which superseded the initial versions on the production line remained one of the principal fighter types defending these vital areas for many years. The upgraded Su-15-98M aerial intercept weapons system comprising this aircraft and the Vozdukh-1M GCI system which permitted guidance in manual, semi-automatic (flight director) and fully automatic modes was capable of intercepting targets flying at speeds of 500-2,500 km/h (310-1,550 mph) and altitudes of 500-24,000 m (1,640-78,740 ft).

The Su-15TM also saw a good deal of action in defence of the Soviet borders, particularly in the late 1970s and the 1980s. 20th April 1978 was the first occasion when a South Korean aircraft 'accidentally' strayed into Soviet airspace. The full truth about this incident remains unknown to this day. Some Western media maintain that the incursion was a result of crew error because the pilots were making their first flight in an unfamiliar aircraft along an unfamiliar route. Get real. It is hard to imagine a navigation error that would lead to a course change in excess of 180°. The facts: at 2054 hrs Moscow time the radar pickets of the 10th Independent PVO Army detected an aircraft 380 km (236 miles) north of Rybachiy Peninsula, flying at 10,000 m (32,800 ft) and heading towards Soviet territorial waters at about 900 km/h (559 mph). When the target approached the 100-km territorial waters strip, at 2111 hrs the officer of the day at the 10th Independent PVO Army headquarters ordered a scramble. Since the unit based nearest to the coast was re-equipping with new aircraft and was not operational for the time being, the mission fell to the 431st IAP at Afrikanda AB (Arkhangel'sk Region), and a Su-15TM piloted by Capt. Aleksandr I. Bosov took off to intercept the target. After being directed towards the unknown aircraft in head-on mode by GCI control the pilot reported seeing the target on his radar display, executed a port turn and started closing in on the target. Coming within visual identification range, Bosov reported it was a four-engined Boeing 747 (*sic*) but he could not make out the insignia – they were either Japanese, Chinese or Korean. (Obviously the pilot had seen hieroglyphic characters of the aircraft's fuselage but had no way of knowing what language it was – *Auth.*)

Actually this was no 747 but a Korean Air Lines Boeing 707-321BA-H registered HL7429 (c/n 19363, fuselage number 623) bound from Paris-Orly to Seoul on flight KE902. As David Gero wrote in his book *Flights of Terror – Aerial Hijack and Sabotage since 1930*, 'Built more than a decade earlier, the aircraft lacked a modern inertial navigation system, and as a magnetic compass is useless in this part of the world* (it gives false readings due to the proximity of the North Pole – *Auth.*), *and with a scarcity in ground aids, the crew would have to rely upon the older but well-proven method of celestial navigation.*

Su-15TM '06 Blue' taxies out with a full complement of missiles – an R-98T to port, and R-98R to starboard and two R-60Ms.

This page:
A fine landing study of a Su-15TM returning from a practice sortie (note the empty pylons) with the flaps fully deployed.

A Su-15TM toting four missiles streams wingtip vortices as it makes a banking turn.

Su-15TM '15 Yellow' seen over international waters with a complement of two R-98s and two cannon pods.

Su-15TM '26 Orange' inspects a Lockheed P-3 Orion over international waters.

Opposite page:
Another *Flagon-F*, '38 Yellow', pictured during a routine sortie.

Su-15TM '04 Red' streams its brake parachute as it lands at an airbase in the Soviet Far East, with typically rugged scenery as a backdrop.

When the MiG-31 entered service, some of the Su-15TMs were transferred to the Air Force's Tactical Aviation branch and painted in a tactical camouflage scheme. The camouflage on '12 Yellow' is already flaking. Note that the skin around the engine nozzles has been left unpainted.

THE Su-15 IN ACTION

Opposite page:
Su-15UM '61 Blue' on a flight line equipped with centralised power and fuel supply (note the outlets between each pair of parking spots). Oddly, sister ship '62 Blue' has the code in a different typeface, despite belonging to the same regiment.

This page:
Atmospheric shots of a Su-15TM silhouetted against the darkening sky as it flies over heavy clouds.

*Trouble first arose in the vicinity of Iceland, when atmospheric conditions prevented the aircraft from communicating with the corresponding ground station. Approximately over Greenland, and **following the instructions of the navigator**, the 707 **inexplicably** initiated a turn of 112 degrees, heading in a south-easterly direction towards the USSR (our highlighting – Auth.). A while later the pilot, Captain Kim Chang Kyu, sensed something was amiss by the rather obvious fact that the sun was on the wrong side of the aircraft!'*

Capt. Bosov was instructed to force the intruder down at a Soviet airfield, which he tried to do, making two passes along the 707's port side 50-60 m (165-200 ft) away to a point ahead of the flight deck and rocking the wings. Yet the South Korean crew ignored these 'amorous advances'.

Meanwhile, after analysing the target's track plotted by AD radars, the 10th Independent PVO Army HQ decided the 707 was pressing on towards the Finnish border, which was only five minutes away, in an attempt to escape and ordered the airliner shot down. At 2142 hrs the fighter pilot fired a single R-98MR missile, reporting an explosion and saying that the target was losing altitude; Bosov was about to fire a second missile but lost target lock-on because the Boeing was descending sharply.

In the meantime a steady exchange of information was going on between PVO command centres at all echelons. The PVO Commander-in-Chief was belatedly informed that the target was a civil airliner; hence the C-in-C's order not to shoot the intruder down but to force it down in one piece reached the lower echelons too late when the 707 was already under fire. The explosion tore away the Boeing's port wingtip and aileron, knocked out the No.1 engine and

A Su-15UM taxies out for take-off at a Far Eastern airbase.

apparently punctured the fuselage, causing a decompression. The crew initiated an emergency descent, causing the PVO radar pickets to briefly lose sight of the aircraft. By then, apart from Bosov's Su-15, five other aircraft had scrambled to intercept the intruder – two Yak-28Ps from Monchegorsk, one MiG-25P from Letneozyorsk, a Su-15TM from Poduzhem'ye AB and a further Su-15TM from Afrikanda AB. When the target vanished from the radarscopes, a further Yak-28P, a MiG-25P and three Su-15TMs from the same bases joined the hunt. A 265th IAP Su-15TM from Poduzhem'ye even fired a missile at a slow-flying target at 5,000 m (16,400 ft) – which later turned out to be nothing more than a cloud of honeycomb filler fragments from the 707's damaged wing.

The crippled airliner circled at low altitude near Loukhi settlement near Kem', Arkhangel'sk Region, where it was again detected and tracked by AD radars and the nearest interceptor was directed to the scene. Since the Su-15's radar was not much use against a low-flying target, the pilots had to rely in the Mk 1 eyeball; yet mortal men haven't got the eyesight of an owl, and even on a cloudless polar night it takes time to locate the target. At 2245 hrs Capt. Keferov of the 265th IAP spotted the intruder flying at 800 m (2,620 ft) near Loukhi; twelve minutes later the target was spotted by another 265th IAP pilot, Maj. A. A. Ghenberg. Together they gave signals to the crew, trying to force the jet to follow them; the airliner ignored the signals, landing on the frozen Lake Korpijärvi 5 km (3.1 miles) south-west of Loukhi. Of the 109 occupants, two passengers were killed (allegedly by fragments of the damaged engine) and 13 people were injured. The crew and passengers of the 707 were detained by the Soviet authorities but subsequently released; the airliner, which was declared a write-off, was recovered from the scene and taken to Moscow for examination.

A much more tragic incident with far-reaching political consequences took place on the night of 1st September 1983. Another Korean Air Lines aircraft, Boeing 747-230B HL7442 (c/n 20559, f/n 186) bound from New York City to Seoul via Anchorage on flight KE007 with, strayed from its designated airway R-20, which passes just 28.2 km (17.5 miles) from the Soviet (Russian) border at the closest point, and entered Soviet airspace near the Kamchatka Peninsula. For 2.5 hours the aircraft flew illegally over a piece of Soviet territory packed with sensitive military installations, and of course it was immediately assumed to be a spyplane, and orders were given to intercept the intruder.

First, a MiG-23 scrambled from Petropavlovsk-Kamchatskiy/Yelizovo airport; it caught up with the target but soon had to give up the chase after running low on fuel. The reason was that after Belenko's defection the top command distrusted the 'grass roots', and the fighters were filled up with *just so much fuel as to make a defection impossible*! No one seemed to realise, or care, that this jeopardised the PVO units' ability to fulfil their mission and that one bad egg does not automatically mean that the whole box of eggs is bad.

Anyway, the 747 left Soviet airspace for a while. However, as it continued in a straight line over the Sea of Okhotsk its course was bound to take it into Soviet airspace again over Sakhalin Island. By then the Soviet PVO system was in turmoil and orders had been issued to shoot the aircraft down, should it intrude again – which it did at 0616 hrs local time. At 0542 hrs and 0554 hrs two 777th IAP Su-15s scrambled from Sokol AB near Dolinsk in the south of Sakhalin Island; the fighters were armed with a pair of R-98 missiles and a pair of UPK-23-250 cannon pods each. One of the Su-15s, with Maj. Ghennadiy N. Osipovich at the controls, intercepted the airliner, which was cruising at 11,000 m (36,090 ft). Osipovich tried to contact the crew by radio and fired warning shots from his cannons, ordering it to land. However, the cannon shells were not tracers, and the Korean crew failed to notice them.

Neither the officer at the PVO command centre nor the pilot was able to identify the intruder aircraft because the incident took place at night. (However, even in poor lighting conditions the Boeing 747 is easy to identify by its unmistakable humpbacked silhouette.)

Since the intruder ignored all calls and pressed on towards the border, orders were given to destroy it. At 0626 hrs Osipovich fired both missiles, which found their mark, damaging the hydraulics and the control system (contrary to some reports, the 747 did not break up in mid-air). After climbing briefly to 11,600 m (38,060 ft) the aircraft suffered a complete decompression and began a spiral descent; at 0638 hrs the jet vanished from the radarscopes at 5,000 m (16,400 ft). Moments later it plunged into the Sea of Japan off Moneron Island, disintegrating on impact and killing all 246 passengers and 23 crew. The incident provoked a huge public outcry and a hysterical anti-Soviet campaign led by the USA.

Even today it is not clear what the KAL jumbo was doing for 2.5 hours in a place where it should not have been at all. Was it on a premeditated spy mission, as the Soviet government claimed, or was the incursion a result of a navigation error? There are several possible explanations and facts to support both theories; however, this is a major topic which lies outside the scope of this book.

Of course, such incidents involving civil airliners do not speak volumes for the Su-15's virtues as an interceptor, being only pages in its service career. Still, according to Soviet fighter pilots' recollections, the reconnaissance aircraft of the 'potential adversary' took pains to avoid coming within the Su-15's reach. A notable exception is the Lockheed SR-71A Blackbird spyplane capable of Mach 3 flight; the MiG-25 was the only Soviet aircraft which was a match for the Blackbird. Of course, the presence of these very different interceptors in the PVO inventory (not counting the Tupolev Tu-128 heavy interceptor, the MiG-23P etc.) increased the overall efficiency of the nation's air defences. The potent MiG-25P was a complicated aircraft to build due to its welded steel airframe and its operations were hampered by the scarcity of its R15B-300 engines which were also used by the many reconnaissance/strike versions of the *Foxbat*; conversely, the less capable Su-15 was easy to build and well adapted for mass production as far as airframe, powerplant and equipment were concerned.

A major problem which the Soviet PVO had to deal with was the large number of drifting reconnaissance balloons launched from Western Europe. Frequently, despite all efforts to destroy it, such a balloon would pass over the entire country. Quite apart from their reconnaissance mission, the dastardly balloons presented a serious danger of collision for civil and military aircraft alike. Supersonic interceptors had limited success in combating reconnaissance balloons, primarily because the target usually had a very small RCS; the aircraft's radar could only detect them at close range, which left very little time for an attack.

Su-15 pilots started their score in the autumn of 1974 when the first balloons were shot down. On 17th October PVO radar pickets detected Yet Another Evil Balloon drifting at 13,000 m (42,650 ft) over the Black Sea and about to enter Soviet airspace. Three 62nd IAP Su-15s took off from Bel'bek AB, making consecutive firing passes at the target; the last of the three managed to shoot off the balloon's reconnaissance systems pod with an R-98T missile. By far the greatest number of such sorties was flown in 1975 – and it was the most successful year as well, 13 out of 16 balloons being destroyed, including five downed by Su-15s.

It should be noted that most of the intruders the Su-15 had to deal with were anything but the *typical targets* it had been designed to intercept. As often as not the target was a light aircraft which was no easy target for a supersonic interceptor due to the huge difference in airspeeds. To add offence to injury, the intruding light aircraft usually flew at ultra-low level where the interceptor's radar could not get a lock-on; this meant the target had to be located visually, and the

Su-15UM '30 Blue' is pictured as it becomes airborne on a practice sortie.

Above: After the dissolution of the Soviet Union part of the Su-15 fleet was taken over by the newly-independent Ukraine. Here a Ukrainian Air Force Su-15TM is seen at Bel'bek AB.

Below: A camouflaged Su-15TM seen in a revetment at Bel'bek AB in the early 1990s. Note the red star on the tail (not yet replaced by the UAF shield-and-trident insignia) and the old code '33 Yellow' bleeding through the current code '23 White'.

Opposite, top: Ukrainian Air Force Su-15TM '28 Blue' and Su-15UM '60 Blue' are readied for a sortie on the flight line at Bel'bek AB.

Opposite, bottom: Ukrainian Air Force Su-15UM '60 Blue' taxies out for take-off. The aircraft looks well-maintained.

view from the Su-15's cockpit left a lot to be desired. This was when accurate guidance by GCI centres proved crucial.

One of the first such incidents occurred on 21st June 1973. At 0836 hrs local time a radar picket of the Baku PVO District detected a target over Iranian territory 300 km (186 miles) south-east of Baku, moving towards the Soviet border at 2,000 m (6,560 ft). Five minutes later a Su-15 took off to ward off the potential intruder; it was soon joined by two more Su-15s of the 976th IAP which was temporarily redeployed to Nasos-

naya AB near Baku due to runway resurfacing work at their own base and a quartet of MiG-17PFUs of the 82nd IAP home-based at Nasosnaya AB.

The intruder crossed the Soviet border at 0859 hrs near the so-called Imishli Salient 170 km (105 miles) south-west of Baku, descending to 200 m (660 ft) to avoid detection by radar. This complicated things considerably for the Su-15 pilots; nevertheless, at 0909 hrs the aircraft, a twin-engine Rockwell Aero Commander, was detected and hemmed in, making an involuntary landing at Nasosnaya AB 27 minutes later. It transpired that the pilot and the sole passenger were heading from Tabriz to the small borderside town of Parsaabad but had lost their way in the mountains. Well, well…

It was no success story on 25th July 1976 when a 'visiting' Cessna 150 Aerobat got away. At 1913 hrs the low-flying intruder was visually detected by border guards troops on the ground, as the PVO radar pickets had missed it. At 1927 hrs a 431st IAP Su-15TM piloted by Capt. Vdovin took off from Afrikanda AB. Nevertheless, the

Ukrainian Air Force Su-15TM '28 Blue' comes in to land at Bel'bek AB, showing the very faded insignia on the tail.

Cessna insolently landed at the PVO reserve airfield at Alakurtti which was conveniently close at hand, the crew refuelled the aircraft, using a spare can of petrol, and continued on their eastward quest unhindered.

Approximately at 1950 hrs the GCI centre directed the Su-15 towards the intruder (which had not avoided detection altogether). Due to poor weather Vdovin was forced to fly below the clouds; still, he managed to spot the Cessna but then lost it from sight and could not regain contact. Two more Su-15TMs and a UTI-MiG-15 trainer (!) were never even directed towards the target. Thus the Finnish-registered Cessna flew on for another 300 km (186 miles) into the depths of the Karelian ASSR but then came to grief, flipping over on its back during a forced landing in a clearing in the woods. Soon afterwards the local residents found the crew and made a 'citizen's arrest'; the Finns claimed they had 'lost their bearings'.

A huge scandal erupted within the PVO system. The PVO C-in-C issued an order requiring that the pilots' gunnery training be stepped up; also, to ensure interception of low- and slow-flying targets like this one the QRA flights of Su-15 units was to include an aircraft armed with UPK-23-250 cannon pods by all means. As a result, from 1970 onwards the aircraft in a QRA flight were armed differently (for example, the flight leader carried two missiles (an R-98TM and an R-98RM) and two drop tanks while the wingman had the same complement of AAMs plus two cannon pods.

The mid-1980s saw a dramatic increase in the requirements which modern interceptors had to meet; new long-range AAMs and more capable aircraft to carry them were developed. Thus the Su-15 was relegated to second place in the PVO inventory, making way for such aircraft as the world-famous Su-27. Some of the Su-15s were transferred to the Soviet Air Force's tactical arm (FA – *Fronto**va**ya avi**ah**tsiya*), exchanging their natural metal finish for a green/brown tactical camouflage. However, the Su-15 was obviously no good as a strike aircraft, since it lacked the appropriate targeting equipment; actual operations soon confirmed this and the type did not gain wide use with the FA.

Like its precursors, the Su-15 was never exported; however, it did see overseas deployment. The 54th GvIAP deployed to Poland about once in every two years to practice operations from stretches of highway used as tactical strips; nothing of the sort existed in the USSR. For instance, in the summer of 1975 a squadron of the 54th GvIAP (by then equipped with the Su-15TM) was on temporary deployment at Słupsk.

The demise of the Soviet Union brought an end to the Su-15's service career in Russia. Even aircraft with plenty of airframe life remaining were struck off charge and scrapped in keeping with the Conventional Forces in Europe (CFE) limitation treaty. The Ukraine hung on to its Su-15s a little longer; the last of the type remained in service with the 636th IAP at Kramatorsk and the 62nd IAP at Bel'bek until 1996.

The Su-15 logically completed the line of Sukhoi's delta-winged interceptors that started with the Su-9, and its withdrawal was a bit hasty since the aircraft still had development and upgrade potential. It may have benefited from the installation of a new radar with 'look-down/shoot-down' capability, for instance. As a result, in 1976 the PVO fighter units started converting *en masse* to the MiG-23M which had this capability. This aircraft was, in turn, succeeded by the MiG-31M and the Su-27P representing a new generation of interceptor technology.

THE Su-15 IN ACTION 55

Above: The end of the road. After retirement from Air Force service these Su-15TMs were stored for a while at Novosibirsk.

Left: Wearing a new coat of paint and the code '11 Red', the T-58L is on display at the Central Russian Air Force Museum in Monino.

A Su-15 *sans suffixe* on display in the now-defunct open-air museum at Moscow-Khodynka in the mid-1990s. The radome has been detached to expose the antenna dish of the Oryol-D58 radar.

Su-15UT '50 Red' and Su-15 '85 Red' were likewise displayed in the museum at Khodynka.

Su-15TM '16 Blue' is preserved in the State Aviation Museum at Kiev-Zhulyany airport.

The Su-15 in Detail

The Su-15 is a single-seat supersonic interceptor designed for day and night VMC/IMC operations. The all-metal airframe is made mostly of D-16 duralumin, V95 and AK4-1 aluminium alloys; highly stressed parts are made of 30KhGSA, 30KhGSNA and 30KhGSL grade steel, while parts of the rear fuselage structure subjected to high temperatures are made of OT4- titanium alloy.

The fuselage is a semi-monocoque riveted stressed-skin structure whose cross-section changes from circular (in the forward fuselage) to almost rectangular (in the area of the air intakes) to elliptical with the longer axis horizontal (in the rear fuselage). It is built in two sections, with a break point at frames 34/35 allowing the rear fuselage to be detached for engine maintenance and removal; the two fuselage sections are held together by bolts. The *forward fuselage* (Section F-1) includes the radome, the pressurised cockpit, two avionics/equipment bays fore and aft of it, the nosewheel well under the cockpit, the air intake assemblies and inlet ducts, the engine bays and the fuel tank bays. The detachable GRP radome tipped with the main air data boom has a simple conical shape on the Su-15/Su-15T/Su-15U and an ogival shape on the Su-15TM/Su-15UT.

The forward avionics bay (frames 1-4) houses the radar set. On single-seat versions the cockpit is contained by pressure bulkheads at frames 4 and 14A and enclosed by a two-piece teardrop canopy. The fixed windshield features curved triangular Perspex sidelights and an elliptical optically-flat bulletproof windscreen made of silicate glass; the aft-sliding canopy portion with blown Perspex glazing can be jettisoned manually or pyrotechnically in an emergency. The Su-15UT trainer features a longer forward fuselage; the tandem cockpits enclosed by a common four-piece canopy with individual aft-hinged sections over the two seats. In contrast, the Su-15UM trainer does not have a fuselage stretch, being dimensionally identical to the single-seat Su-15TM.

The forward fuselage is area-ruled, starting at the cockpit section which is flanked by the air intakes. The air intake trunks are rectangular-section structures blending into the centre portion of the fuselage. The intakes have sharp lips and are provided with boundary layer splitter plates which are attached to the fuselage by V-shaped fairings spilling the boundary layer. To improve performance at high angles of attack the air intake trunks are canted outward 2°30'. Each intake features a three-segment vertical airflow control ramp and a rectangular auxiliary blow-in door on the outer side; these are governed by the UVD-58M engine/intake control system. The centre portion (frames 14A-34) accommodates three integral fuel tanks, the inlet ducts, whose cross-section gradually changes to circular, and the engine bays (frames 28-34); it features two 'wet' external stores hardpoints side by side. Titanium firewalls and

The forward fuselage of a Su-15 *sans suffixe*, showing the detachable panels for maintenance access to the radar set. Note that the radome has a metal attachment 'skirt' and that the joint line of the dielectric part is not at right angles to the fuselage waterline.

The canopy of an early Su-15 lacking the rear view mirror. The inscription reads 'Keep the canopy covered when parked'.

Another view of Su-15's canopy, showing the electric de-icing threads on the sliding portion. Note the Sukhoi OKB's unofficial 'winged archer' badge, with the Guards badge (right) and the Order of the Red Banner of Combat painted below it.

The cockpit, nose gear unit and starboard air intake of a production Su-15UT. Note the shape of the anti-glare panels, the forward-vision periscope built into the instructor's canopy section and the V-shaped rain deflector ahead of the windscreen.

This 'toad's eye view' of a Su-15TM shows the canted air intakes with boundary layer splitter plates standing proud of the fuselage. Note how the fuselage is area-ruled in the cockpit area.

heat shields are provided in the engine bays to contain possible fires. The *rear fuselage* (Section F-2) carries the tail surfaces and a detachable 'pen nib' fairing made of titanium and stainless steel sheet between the engine nozzles; it accommodates the engine jetpipes and the tailplane actuators. Four airbrakes with a total area of 1.32 m² (14.19 sq ft) are located between frames 35-38, opening 50°.

The cantilever mid-set wings are of simple delta shape with 60° leading-edge sweep on the Su-15 *sans suffixe*, with 2° anhedral and zero incidence. The Su-15T and Su-15TM have increased-area wings with a leading-edge kink at 2.625 m (8 ft 7⅜ in) from the centreline, the outer portions featuring 45° leading-edge sweep and 7° negative camber. The wings are one-piece structures of two-spar stressed-skin construction joined to the fuselage at mainframes 16, 21, 25, 28 and 29. On the pure delta version each wing has 17 ribs, 28 rib caps and three auxiliary spars which, together with the front and rear spars, form five bays: the leading edge, forward bay, mainwheel well (between the Nos. 1 and 2 auxiliary spars), rear bay and trailing edge. The double-delta versions have 18 ribs and 29 rib caps per wing. The bays between the Nos. 2 and 3 auxiliary spars are integral fuel tanks whose skin panels are stamped integrally with the ribs and stringers; ordinary sheet metal skins are used elsewhere. Each wing has a single boundary layer fence on the upper side. The wings have one-piece blown flaps, with one-piece ailerons outboard of these. The flaps are hydraulically actuated, with pneumatic extension in an emergency; flap settings are 15° for take-off and 25° for landing when the BLCS is on, or 25° and 45° respectively with the BLCS off. The ailerons are aerodynamically balanced and mass-balanced, with a travel limit of ±18°30'. There are two external stores hardpoints on each wing.

The conventional tail surfaces swept back 55° at quarter-chord are of riveted stressed-skin construction. The *vertical tail* comprises a one-piece fin an inset rudder. The fin is a single-spar structure attached to fuselage mainframes 35 and 42, with a rear auxiliary spar (internal brace), front and rear false spars, stringers and ribs; it features a root fillet and a glassfibre tip fairing. The mass-balanced rudder is a single-spar structure; it is carried on three brackets. The cantilever *horizontal tail* mounted 110 mm (4⅜ in) below the fuselage waterline consists of differentially movable slab stabilisers (stabilators) turning on stub axles attached to fuselage mainframe 43; anhedral 2°, incidence in neutral position –4°10'. Each stabilator is a single-spar structure with front and rear false spars, stringers and ribs, and sheet duralumin skins. The stabilators feature anti-flutter booms angled 15° upwards at the tips.

The tricycle landing gear is hydraulically retractable, with pneumatic emergency extension; all units have levered suspension and oleo-pneumatic shock absorbers. The forward-retracting nose unit is equipped with a shimmy damper, castoring through ±60°; steering on the ground is by differential braking. On the Su-15 *sans suffixe* the nose unit has a single 660 x 200 mm (26.0 x

The starboard engine's air intake. The hinged airflow control ramp is just visible. The inscriptions between the black/yellow 'tiger stripes' on the boundary layer splitter plate read *Opasno, vozdukhozabornik* (Danger, air intake).

7.87 in) KT-61/3 wheel (*koleso tormoznoye* – brake wheel). The Su-15T, Su-15TM and Su-15UM have a taller nose gear unit with twin 620 x 180 mm (24.4 x 7.0 in) KN-9 wheels (*koleso netormoznoye* – non-braking wheel). The main units retracting inward into the wing roots are equipped with 880 x 230 mm (34.6 x 9.0 in) KT-117 brake wheels on all versions. All brake wheels feature pneumatically actuated disc brakes. The nosewheel well is closed by clamshell doors, the mainwheel wells by triple doors (one segment is hinged to the front spar, one to the root rib and a third segment attached to the strut); all doors remain open when the gear is down. A PT-15 ribbon-type brake parachute with an area of 25 m² (268.8 sq ft) housed in a fairing at the base of the fin is provided to shorten the landing run.

The Su-15 and Su-15UT are powered by two Tumanskiy R11F2S-300 or R11F2SU-300 axial-flow afterburning turbojets rated at 3,900 kgp (8,600 lbst) at full military power and 6,175 kgp (13,610 lbst) in full afterburner. The R11F2S-300 is a two-spool turbojet featuring a three-stage low-pressure (LP) compressor, a three-stage high-pressure (HP) compressor, an annular combustion chamber, single-stage HP and LP turbines and an afterburner with an all-mode variable nozzle. The transonic-flow compressor has no inlet guide vanes; the second-stage blades have snubbers to prevent resonance vibrations. Bleed valves are provided; the R-11F2SU-300 version features air ducts for the BLCS, with a maximum bleed rate of 2.5 kg/sec (5.5 lb/sec) in dry thrust mode only. The accessory gearbox is located ventrally. The engine has a closed-type lubrication system. Starting is electric (by means of a starter-generator), using onboard or ground DC power. Engine pressure ratio at take-off thrust 8.7; mass flow at take-off thrust 66 kg/sec (145.5 lb/sec), maximum turbine temperature 1,175° K. SFC 2.37 kg/kgp·h (lb/lbst·h) in full afterburner and 0.93 kg/kgp·h in cruise mode. Length overall (including afterburner) 4,600 mm (15 ft 1 in), casing diameter 825 mm (2 ft 8½ in); dry weight 1,088 kg (2,400 lb).

The port wing of a Su-15TM, showing the flap and the aileron terminating short of the wingtip. Note how the boundary layer fence is built in three pieces to cater for the wing flexure in flight.

The vertical tail of a Su-15 *sans suffixe*.

The vertical tail of a Su-15TM, showing the radar warning receiver antenna pod relocated to a position above the brake parachute fairing and the rudder cut away to accommodate it.

The port stabilator, showing the upward-angled anti-flutter boom and the slit aft of the pivot for the actuator arm.

The nose gear unit of a Su-15TM with twin KN-9 wheels. Note the appropriately bulged nosewheel well doors.

The port main gear unit of a Su-15. One of the door segments is missing.

The Su-15T, Su-15TM and Su-15UM are powered by two Tumanskiy R13-300 axial-flow afterburning turbojets rated at 4,100 kgp (9,040 lbst) at full military power and 6,600 kgp (14,550 lbst) in full afterburner. This is likewise a two-spool turbojet with a three-stage LP compressor, a five-stage HP compressor, an annular combustion chamber, single-stage HP and LP turbines and an afterburner with an all-mode variable nozzle. The afterburner features annular/radial flame holders and a perforated heat shield. EPR 9.15 at full military power and 9.25 in full afterburner; mass flow at take-off thrust 66 kg/sec, maximum turbine temperature 1,223° K. SFC 2.25 kg/kgp·h in full afterburner and 0.96 kg/kgp·h in cruise mode. Length overall 4,600 mm, casing diameter 907 mm (2 ft 11¾ in); dry weight 1,134 kg (2,500 lb).

The port side of a Su-15's centre fuselage with the maintenance hatches open, showing the accessories and oil tank of the port engine.

Two drop tanks on the fuselage hardpoints of a Su-15.

■ Su-15 SPECIFICATIONS

	Su-15	Su-15TM	Su-15UT	Su-15UM
Powerplant	2 x R11F2S-300	2 x R13-300	2 x R11F2S-300	2 x R13-300
Length (less pitot)	20.54 m (67 ft 4⅝ in)	20.54 m (67 ft 4⅝ in)	20.99 m (68 ft 10⅜ in)	19.66 m (64 ft 6 in)
Wing span	8.616 m (28 ft 3¼ in)	9.34 m (30 ft 7¾ in)	8.616 m (28 ft 3¼ in)	9.34 m (30 ft 7¾ in)
Height on ground	5.0 m (16 ft 4⅞ in)	4.843 m (15 ft 10⅝ in)	5.0 m (16 ft 4⅞ in)	4.843 m (15 ft 10⅝ in)
Landing gear track	4.79 m (15 ft 8½ in)	4.79 m (15 ft 8½ in)	4.79 m (15 ft 8½ in)	4.79 m (15 ft 8½ in)
Landing gear wheelbase	5.887 m (19 ft 3¾ in)	5.942 m (19 ft 6 in)	n.a.	5.942 m (19 ft 6 in)
Wing area, m^2 (sq ft)	34.56 (371.6)	36.6 (393.54)	34.56 (371.6)	36.6 (393.54)
Horizontal tail area, m^2 (sq ft)	5.58 (60.0)	5.58 (60.0)	5.58 (60.0)	6.43 (69.1)
Vertical tail area, m^2 (sq ft)	6.951 (74.74)	6.951 (74.74)	6.951 (74.74)	6.951 (74.74)
TOW, kg (lb):				
normal	16,520 (36,420) [2]	17,194 (37,905) [5]	16,690 (36,795) [7]	17,200 (37,920) [2]
maximum	17,094 (37,685) [1]	17,900 (39,460) [4]	17,200 (37,920) [1]	17,900 (39,460) [4]
Empty weight, kg (lb)	10,220 (22,530)	10,874 (23,970)	10,750 (23,700)	10,635 (23,445)
Landing weight, kg (lb)	12,040 (50,790)	12,060 (26,590)	n.a.	13,314 (29,350)
Fuel load, kg (lb)	5,600 (12,345)	5,550 (12,235)	5,010 (11,045)	5,550 (12,235)
Thrust/weight ratio	0.92	0.92	0.88	n.a.
Top speed, km/h (mph):				
at sea level	1,200 (745)	1,300 (807)	1,200 (745)	1,250 (776)
at high altitude	2,230 (1,385) [3]	2,230 (1,385) [6]	1,850 (1,150) [3]	1,875 (1,164) [8]
Mach number at high altitude	2.13	2.16	1.753	1.758
Climb time to 16,000 m (52,490 ft), minutes	13	n.a.	12	n.a.
Service ceiling, m (ft)	18,500 (60,695)	18,500 (60,695)	16,700 (54,790)	15,500 (50,850)
G limit	5	5	5	5
Unstick speed, km/h (mph)	395 (245)	370 (230)	n.a.	340-350 (211-217)
Landing speed, km/h (mph)	315 (195)	285-295 (177-183)	330-340 (204-211)	260-280 (161-174)
Range, km (miles):				
on internal fuel	1,270 (790)	1,380 (860)	1,290 (800)	n.a.
with drop tanks	1,550 (960)	1,700 (1,055)	1,700 (1,055)	1,150 (715)
Endurance	1 hr 54 min	n.a.	n.a.	n.a.
Take-off run, m (ft)	1,100 (3,600)	1,000-1,100 (3,280-3,600)	1,200 (3,940)	n.a.
Landing run, m (ft):				
without brake parachute	1,500 (4,920)	1,050-1,150 (3,440-3,770)	n.a.	n.a.
with brake parachute	1,000 (3,280)	850-950 (2,790-3,120)	1,150-1,200 (3,770-3,940)	n.a.
Armament:				
missiles	2 x R-98R/T (R-8MR/MT, R-8MR1/MT1)	2 x R-98MR/MT (R-8MR/MT, R-8MR1/MT1)	2 x R-8MT 2 x R-60	2 x R-98MT 2 x R-60
cannons	2 x UPK-23-250 [9]	2 x UPK-23-250 [9]	–	–

1. With two drop tanks/no missiles; 2. With two R-98 missiles; 3. At 15,000 m (49,210 ft); 4. With two drop tanks, two R-98Ms and two R-60s;
5. With two R-98 missiles; 6. At 13,000 m (42,650 ft); 7. With two dummy R-98 missiles; 8. At 11,500 (37,730 ft);
9. The UPK-23-250 houses a GSh-23 twin-barrel cannon with 250 rounds

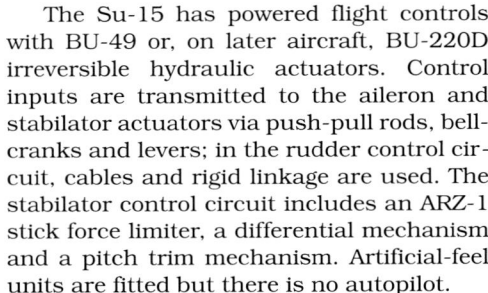

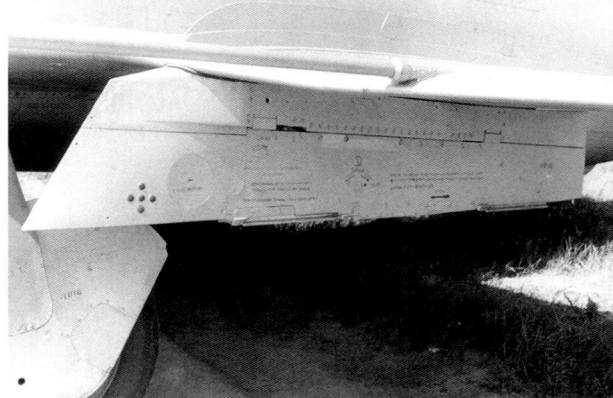

The Su-15 has powered flight controls with BU-49 or, on later aircraft, BU-220D irreversible hydraulic actuators. Control inputs are transmitted to the aileron and stabilator actuators via push-pull rods, bell-cranks and levers; in the rudder control circuit, cables and rigid linkage are used. The stabilator control circuit includes an ARZ-1 stick force limiter, a differential mechanism and a pitch trim mechanism. Artificial-feel units are fitted but there is no autopilot.

Internal fuel is carried in five integral tanks – three fuselage tanks (No.1, frames 14A-18; No.2, frames 18-21; No.3, frames 21-28) and two wing tanks. The total capacity stated in different documents varies from 8,675 to 8,860 litres (1,907 to 1,949 Imp gal). The 'wet' hardpoints under the fuselage permit carriage of two 600-litre (132 Imp gal) drop tanks.

Main 28.5 V DC electric power provided by two 12-kW GSR-ST-12000VT engine-driven starter-generators, with two 15-STsS-45A silver-zinc batteries providing 22.5 V DC as a back-up. 115 V/400 Hz single-phase AC provided by two SGO-8TF engine-driven generators. Two ground power receptacles are provided on the port side of the fuselage. The exterior lighting includes BANO-45 navigation lights at the wingtips, a KhS-39 tail navigation light on the fin trailing edge, and two retractable PRF-4M landing/taxi lights in the wing roots.

There are four hydraulic systems (two primary and two actuator supply systems), each with its own NP-34 or NP-26/1 engine-driven plunger-type pump. The *No.1 primary system* operates the landing gear, flaps, airbrakes, artificial-feel unit switch mechanism, the port air intake ramp and auxiliary blow-in door and the port engine's nozzle actuators. It also performs automatic wheel braking during landing gear retraction The *No.2 primary system* powers the radar dish drive, the starboard air intake ramp/auxiliary blow-in door and the starboard engine's nozzle actuators. The two *actuator supply systems* (main and back-up) power the aileron, rudder and tailplane actuators; the port system features an NS-3 autonomous emergency pump ensuring that the aircraft remains controllable in the event of a dual engine failure. All systems use AMG-10 oil-type hydraulic fluid; nominal pressure 215 kg/cm² (3,070 psi).

The pneumatic system performs normal and emergency wheel braking, emergency landing gear and flap extension, and operates the inflatable canopy perimeter seal. The system is charged with compressed air to 200 bars (2,857 psi), featuring three 6-litre (1.32 Imp gal) air bottles. There is also a separate pneumatic system charged to 150 bars (2,140 psi) which operates the stabilising gyros of the missiles' seeker heads.

The cockpit is pressurised (and the canopy demisted) by air bled from the engine's fifth or seventh compressor stage, depending on rpm. Cockpit air temperature is maintained automatically at +10-20°C (50-68°F) by a TRTVK-45M regulator; the air pressure is governed by an ARD-57V automatic pressure regulator. The oxygen equipment includes a breathing apparatus for normal operation and a separate breathing apparatus used in the event of an ejection. For operations at altitudes up to 10,000 m (32,810 ft) and speeds up to 900 km/h (560 mph) the pilot is equipped with a KM-32 oxygen mask, a ZSh-3 flying helmet (*zaschchitnyy shlem*) and a VK-3 or VK-4 ventilated flying suit (*ventileeruyemy kostyum*). For missions involving supersonic flight the pilot wears a VKK-4M, VKK-6 or VKK-6P pressure suit and a GSh-4MS, GSh-4MP or GSh-6M full-face pressure helmet (*ghermoshlem*).

Fire protection is provided by an SSP-2I fire warning system (*sistema signalizahtsii pozhahra*) and a UBSh-6-1 fire extinguisher bottle with two distribution manifolds around the engines.

The navigation and piloting equipment includes a KSI-5 compass system and an RSP-6 instrument landing system (comprising an ARK-10 ADF, an RV-UM low-range radio altimeter, an MRP-56P marker beacon receiver and SOD-57M DME). An RSBN-5S

The cockpit of a Su-15 *sans suffixe*. The radarscope is mounted centrally above the instrument panel.

Opposite page:
An R-98T missile on the port outboard pylon of a Su-15TM.

The black-striped R-98T on the port outboard pylon of this Su-15TM is an inert training round. The inboard pylon carries two R-60M short-range AAMs on an APU-60-2 launcher.

The port inboard pylon of a Su-15TM with an APU-60 single launch rail for an R-60 AAM.

The outboard wing pylons with BD3-59FK launch rails for medium-range AAMs.

This page:
The cockpit of a Su-15TM, showing subtle differences from the earlier version.

Opposite page:
The box top of the 1:48th scale Collect-Aire Su-15TM kit. The manufacturer's name is mis-spelled as 'Sukohi', and Every Single Word Is Capitalised For Some Reason.

The 1:48th scale HitKit Su-15TM built by Phil 'Bondo' Brandt. The build involved correction of the air intake geometry. A modified MiG-21 resin cockpit tub and instrument panel by Aires were used, with a scratchbuilt radarscope, as were a modified KS-3 resin ejection seat by Squadron; modified Su-27 resin exhausts by KMC; airbrakes, landing main gear parts, pitot, pylons and boarding ladder from an OEZ Letohrad Su-7 kit. The nose gear unit was scratchbuilt, as were the cannon pods and anti-flutter booms. The model was painted with Alclad II, and Soviet insignia from an Aeromaster decal sheet were used.

SHORAN and an SAU-58 (or SAU-58-2) automatic flight control system are fitted. The cockpit instrumentation includes a KUSI-2500 airspeed indicator, VDI-30 altimeter, AGD-1 artificial horizon, EUP-53 turn and bank indicator, AM-10 accelerometer (G load indicator), VAR-300 vertical speed indicator, UKL-2 heading indicator, KI-13 flux gate compass, M-2.5 Mach meter, ARK-10 ADF indicator, critical angle of attack warning system and AChKh clock. Air data is provided by main and back-up PVD-7 pitots; the main pitot is located at the tip of the radome.

Communications equipment comprises an RSIU-5V (R-802) VHF radio (or an R-832M Evkalipt radio on the Su-15TM) and, on trainer versions, an SPU-9 intercom. Single-seat versions have a data link receiver making up part of the Lazoor'-M (ARL-SM) GCI system. The aircraft is equipped with an SRZO-2M Kremniy-2M IFF interrogator/transponder, a Sirena-2 or (Su-15TM) SPO-10 Sirena-3 radar warning receiver. An SARPP-12V-1 flight data recorder (*sistema avtomaticheskoy reghistrahtsii parahmetrov polyota* – automatic flight parameter recording system) is fitted From Batch 11 onwards; the Su-15UM has an MS-61B cockpit voice recorder.

The Su-15 *sans suffixe* is equipped with an RP-15 Oryol-D58 or RP-15M Oryol-D58M fire control radar in a conical radome. The Su-15T features a Taifoon radar in an identical radome, while the Su-15TM has an RP-26 Taifoon-M radar in an ogival radome. No radar is fitted to the trainer versions. A K-10T collimator gunsight is provided. The standard armament of the Su-15 *sans suffixe* consists of two R-98R (SARH) and R-98T (IR-homing) medium-range AAMs carried on wing-mounted pylons with BD3-59FK launch rails. The older R-8MR and R-8MT can also be used, but only when the target is attacked in pursuit mode. The missiles are fired singly or in a salvo with a 0.5-second interval. The Su-15TM is armed with two R-98MR (R-98MT) medium-range AAMs and two R-60 short-range IR-homing AAMs on single APU-60 or twin APU-60-2 launch rails; later the Su-15 *sans suffixe* underwent a similar upgrade, being retrofitted with two extra pylons. The radar-less Su-15UM can carry R-98MT and R-60 AAMs. The fuselage hardpoints can be used for carrying free-fall bombs (Su-15TM only) or UPK-23-250 gun pods, each containing a Gryazev/Shipoonov GSh-23 double-barrelled 23-mm (.90 calibre) cannon with 250 rounds.

The crew rescue system features a Sukhoi KS-4 ejection seat (or seats) permitting safe ejection throughout the flight envelope and on the ground at speeds not less than 140 km/h (87 mph).

The Modeller's Corner

Should you fancy building a model of the Su-15 and start browsing the market for a kit that meets your criteria as to accuracy and ease of build (and does not cause your wallet to scream in agony), you will find that the subject has been unanimously ignored by all the major plastic kit manufacturers. One can only guess why. Perhaps it is because the Su-15, which was never exported, was very seldom seen outside its home country and, for obvious reasons, was ungetatable for the purpose of making accurate scale plans used in model making.

Oh, to be sure, the Su-15 is well represented – as of this writing, there are close to 20 kits in two scales from no fewer than eleven (!) brands, but most of these manufacturers are rather obscure. And certainly only one of these manufacturers uses the traditional (high-pressure) injection moulding technology.

1:48th scale

The US company **Collect-Aire Models** offers a limited edition kit of the Su-15TM (Ref. No.4858) in their Pro-Tech series. It is cast in polyester resin, with white metal parts and a vacuform canopy, and features engraved panel lines; the rudder and flaps are separate parts that can be set in a deflected position.

The Polish manufacturer **HitKit** specialising in vacuform models also offers a kit of the Su-15TM. Actually the kit gives you the option of building the Su-15 *sans suffixe*, Su-15TM or Su-15UM. The principal components are made of fairly thin plastic sheet and feature fairly deep recessed panel lines but these are irregular; the landing gear and some other parts are white metal castings but the casting quality is poor. A major inaccuracy of the Polish kit is that the outer faces of the air intakes are vertical, not canted outwards as they should be, and correcting this involves a lot of 'surgery' which the modeller might not care to undertake. The decal sheet is fairly good but the lighter colours are not opaque enough.

For those modellers who want a more builder-friendly kit, in 2002 the well-known Chinese manufacturer **Trumpeter Models** issued three injection moulded kits at once – the Su-15 *sans suffixe* (Ref. No.02810), Su-15TM (Ref. No.02811) and Su-15UM (Ref. No.02812). In typical Trumpeter fashion, the kits are crisply moulded, with seven sprues in light grey plastic (which is quite hard) and one clear sprue, and have maximum commonality. The Su-15 *sans suffixe* kit consists of 177 parts and the finished model is 470 mm (18½ in) long, with a wing span of 178 mm (7 in). The Su-15TM kit consists of 179 parts and the Su-15UM has 190 parts; in both cases the finished model is 476 mm (18⁴⁷⁄₆₄ in) long, with a wing span of 195 mm (7⁴³⁄₆₄ in). The Su-15UM kit includes decals for a Soviet/Russian Air Force or Ukrainian Air Force aircraft.

Left: The box top of Trumpeter's 1:48th scale Su-15 *sans suffixe* kit.

Below: The 1:48th scale Trumpeter Su-15 *sans suffixe* built by Giannis Giavasis from Greece. The model represents an updated aircraft with two extra missile pylons and R-60M AAMs. Photo-etched cockpit and exterior details (including airbrakes) by Eduard Models and cooling air scoops by QuickBoost were used; it was painted with Alclad II paints, with post-shading made by using smoke; panel lines were airbrushed on and weathering was done by paint chipping.

Opposite page: The 1:48th scale Trumpeter Su-15 *sans suffixe* built out of the box by a Russian modeller with the internet alias Lysyy Maks ('Bald Max'). It is painted with Mr. Hobby and Tamiya enamels, highlighted with MiG wash and given a coat of Tamiya varnish.

Left: The box top of Trumpeter's 1:48th scale Su-15TM kit.

Below: The 1:48th scale Trumpeter Su-15TM built by Jon Bryon. The model was built, using a replacement resin radome, resin/PE air data boom and cooling air scoops by QuickBoost; the airbrakes were scratchbuilt. Mr. Colour and Alclad II paints were used, with a coat of Johnson's Klear varnish on some areas, and Begemot decals.

Opposite page:
Here, for comparison, is the 1:48th scale Trumpeter Su-15TM built by Charles Leung. The modeller made an error, believing that the Su-15's radome swings open (in reality, it is detachable).

The main components have finely engraved panel lines; the parts fit together well (except the wing/fuselage joints) and assembly is quite straightforward. On the down side, the Trumpeter brand is not renowned for accuracy, and the Su-15 kits are appallingly inaccurate, being based on very flawed scale plans. Firstly, the axis of the radome is parallel to the fuselage waterline, whereas on the actual aircraft the radome is visibly drooped. On the Su-15TM and Su-15UM the forward fuselage section

The T-58VD converted from a 1:48th scale Trumpeter Su-15TM by Armin Knes.

ahead of the cockpit is a staggering 12 mm (0¹⁵⁄₃₂ in) longer than it should be; besides, the radome shape and the canopy shape are wrong. The rear fuselage is also 3 mm (≈ 0⅛ in) longer than it should be, its aft extremity is too narrow and has an incorrect cross-section, and the airbrakes are too narrow. Secondly, while the wing span of the Su-15TM and Su-15UM is correct, the planform is not: the leading-edge sweep of the inboard portions is in excess of the required 60° – which has caused the centre fuselage near the wing roots to be 5 mm (0¹³⁄₆₄ in) longer than it should be – and the outer wing camber is wrong; besides, the wing anhedral is excessive. The landing lights are located on the underside of the wings – which is correct for the Su-15 *sans suffixe* but not for the Su-15TM/Su-15UM, where they were on the underside of the air intake trunks. Thirdly, the vertical tail is too small and has insufficient leading-edge sweep. The stabilators have the correct span and leading-edge sweep but the wrong planform, with excessive taper and insufficient trailing-edge sweep, and again exaggerated anhedral; moreover, the anti-flutter booms at the tips are 0.4 cm (≈ 0⁵⁄₃₂ in) too short and straight instead of being properly angled upward to avoid scraping the runway on rotation. Fourth, the engraved surface detail is largely useless because it is incorrect – the panel lines on the upper and lower surfaces of the wings don't even meet at the leading edge. Fifth, the nose gear strut is 5 mm too short, resulting in an incorrect ground angle; the outer main gear doors have the wrong shape and position with regard to the struts. The auxiliary air intakes on the fuselage are exaggerated, the missile's proportions and fin shape are not entirely correct and so on. In a nutshell, the Trumpeter kits require extensive plastic surgery (this time the pun is intentional) if the end result is to resemble a Su-15, not a parody of it.

The box top of Trumpeter's 1:48th scale Su-15UM kit.

The 1:48th scale Trumpeter Su-15UM built by Hans Jürgen Bauer. The rear cockpit canopy was mistakenly modelled as aft-sliding; in reality it is aft-hinged.

1:72nd scale

The box top of Pioneer 2's 1:72nd scale Su-15TM ('Su-21F Flagon').

In 1:72nd scale, the Su-15 was first offered back in 1984 by the US manufacturer **Archer's Products, Inc.** (doing business as **Nova Models**), which released a vacuform kit marketed as 'Sukhoi Su-15VD Flagon-F' (*sic*). The hybrid designation may be invented (as mentioned earlier, the V/STOL technology demonstrator was called T-58VD), but the NATO reporting name is correct – the drawing on the instruction sheet shows the ogival radome and cranked-delta wings typical of the standard Su-15TM. The kit includes parts to build the R-98 missiles and the UPK-23-250 cannon pods.

In the early/mid-1980s the British company **I.D. Models**, which specialised in vacuform models, also offered a kit of the Su-15TM, which is clearly the most popular version with the kit manufacturers; the

The Pioneer 2 Su-15TM built by Rodrigo Rejas from Chile. The model was built out of the box and painted with Humbrol enamels; homemade decals were used instead of the sorely inadequate stock ones, and black chalk powder was brushed on for weathering.

Scalemates website (which suggests the model is a rebag – the term 'rebox' is not applicable in this case – of the Nova Models kit) says a 'new box' version appeared in 1985, which should give some idea. The model (Ref. No.78) was *extremely* basic, as many of the I.D. Models kits were. The clear plastic bag (there was no proper box) contained a single sheet of white plastic card with the basic airframe components – a horizontally split (!) fuselage, the wings, vertical tail and stabilators (each in two halves) – and an instruction leaflet. And that's it! There wasn't even a clear acrylic sheet for the cockpit canopy – the latter item was just an opaque bulge on the same white plastic sheet. Well, having to browse the spares box for wheels and such when building vacuform kits is normal, but this excuse for a canopy is simply over the top. In the early 2000s I.D. Models went out of business, selling most of the moulds to Tigger Models who are, as one commentator put it, 'just sitting on them'. Theoretically, the old Su-15 kit may still be available – but who would fancy it now?

In 1986 the US manufacturer **Leoman** based in Glendale, California, released a limited edition injection moulded kit of the Su-15TM cast in polyester resin and polystyrene. The original black/white box (marked 'Temporary package') was titled 'Su-15, Flagon G' – even though the box art showed a single-seat Su-15TM *Flagon-F*, not a trainer – and the box top featured brief specifications of the aircraft. Unfortunately no information is available on the kit itself.

In the 1990s the Turkish company **Pioneer 2** brought out what was probably the first all-plastic injection moulded kits of the subject – apparently using the 'short run' technology. Two versions are represented – the Su-15TM, which is marketed under the erroneous designation 'Sukhoi Su 21F Flagon' (a cross between the erroneous designation initially attributed to the type in the West and its NATO reporting name; Ref. No.5002), and the Su-15UM, which is similarly mis-labelled 'Sukhoi Su 21G Flagon' (Ref. No.5005). The kits are moulded in light blue plastic; in both cases the box top features a properly credited illustration by Don Greer, courtesy of Squadron/Signal Publications (from the 'Su-15 in Action' brochure). Unsurprisingly, the kits are rather crude and accuracy is poor – the model is about one scale foot short in both length and wing span; also, the distinctive droop of the radome has been omitted, the brake parachute container is placed too high and the wing planform is dubious. Curiously, the manufacturer misinterpreted the distinctive black stripes of the 'Danger, air intake' stencils on the boundary layer splitter plates as vertical slits – and dutifully 'reproduced' them as such. The decal sheet is also poor, with nothing but six red stars and a pair of tactical codes. Later, in the 2000s, both kits were rebranded **PM Model**, with the same erroneous designations and the same box art ('Sukhoi Su 21F Flagon', Ref. No.PM-401; 'Sukhoi Su 21G Flagon', Ref. No.PM-402). The company has since gone out of business.

In the 2000s the Russian company **VES Model** (aka **NPO VES**) from Voronezh released a short-run injection moulded kit of the Su-15TM (Ref. No.12). There are two sprues cast in soft dark grey polystyrene with 105 parts, plus a clear sprue with six parts (the canopy can be modelled open or closed); the rudder, flaps and ailerons are separate parts; external stores supplied with the kit are two R-98 AAMs, two R-60 AAMs and two drop tanks. The decal sheet by Begemot Decals features tactical codes for five Soviet/Russian Air Force aircraft, some of the profuse maintenance stencils and even the 13-digit construction numbers for the R-98s. In 2005 the kit was reboxed by another Russian company, **Gran' Ltd.** (Ref. No.7241), with new decals (likewise by Begemot Decals) giving you the option of building a Soviet/Russian Air Force or Ukrainian Air Force aircraft.

The 'gooda news' is that the VES/Gran' kit is one of the most geometrically accurate kits of the *Flagon-F* to date – which is hardly surprising, as the model makers had access to the real thing. The 'bada news' is that, quite apart from the rather poor quality of

The box top of the VES Model 1:72nd scale Su-15TM.

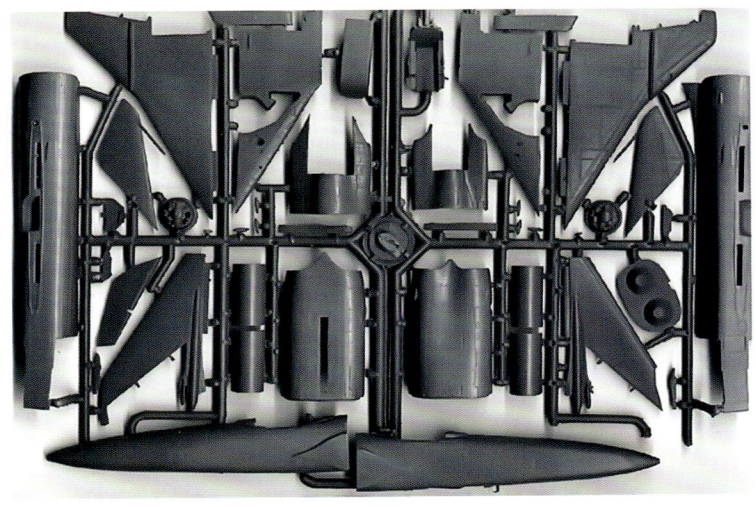

One of the sprues of the VES kit – just to give an idea of how 'exploded' the model is!

The 1:72nd scale Trumpeter Su-15 *sans suffixe* built out of the box by Malik Aminov. The model was painted with Tamiya and Motip paints.

the castings (an inevitable result of the 'short run' technology), the kit is horrendously difficult to build because of the parts breakdown. Consider this: the fuselage comes in eight (!) parts, including the horizontally split rear fuselage section with the correctly positioned fuselage break point (this gives you the option of building the model with the rear fuselage-cum-tail unit detached to expose the engine jetpipes – although the ground handling dolly will have to be scratchbuilt), and the parts are a poor fit. While the panel lines are generally engraved, this is inconsistent, and some are actually raised. The missiles have separate fins, turning assembly into a problem; the ejection seat (described as 'less than stellar') comes in three parts and so on. Also, the two-piece canopy is of uneven thickness, distorting the view. Separating the parts (especially the small ones) from the sprues without damaging them is something of a problem. On the other hand, the size A4 instruction sheet is quite explicit and includes a summary of the Su-15's development history.

True to form, **Trumpeter Models** let loose with a similar salvo of three *Flagon* versions in 2008 – the Su-15 *sans suffixe* (Ref. No.01624), Su-15TM (Ref. No.01623) and Su-15UM (Ref. No.01625). These are scaled-down versions of their 1:48th scale kits described above and consequently suffer from the same inaccuracies. Each kit consists of 100 parts on six sprues, the assembled Su-15 *sans suffixe* having a length of 320.24 mm (12^{39}/$_{64}$ in) and a wing span of 118.73 mm (4^{43}/$_{64}$ in); with the Su-15TM and Su-15UM it is 318.24 mm (12^{17}/$_{32}$ in) and 129.8 mm (5^{7}/$_{64}$ in).

THE MODELLER'S CORNER **77**

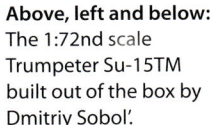

Top left: The box top of Trumpeter's 1:72nd scale Su-15 *sans suffixe* kit.

Top right: The box top of Trumpeter's 1:72nd scale Su-15TM.

Above, left and below: The 1:72nd scale Trumpeter Su-15TM built out of the box by Dmitriy Sobol'.

The Ukrainian manufacturer **Amodel**, which has made a name for itself by catering for modellers who want something outside the mainstream, offers two versions of the *Flagon*, likewise using the 'short run' technology – the Su-15TM (Ref. No.7263) and the Su-15UM (Ref. No.72107) released in 2003. As far as the injection moulded Su-15 kits are concerned, accuracy seems to be in reverse proportion to ease of build, and Amodel's offerings appear to be somewhere in between VES and Trumpeter, which are antipodes in this respect. At any rate, they are suitable for the average modeller whose nerves are not made of steel.

The Su-15TM kit consists of 120 parts arranged on five sprues moulded in a brittle light grey plastic and one clear sprue; the thick canopy is the only transparency, no landing lights being included. The moulding quality is decent as far as short-run kits go, with finely engraved panel lines and very little flash, but some parts show a bit of warping and sink holes. The fuselage is split vertically into two almost full-length parts, with the radome (incidentally, also split lengthwise), air intakes and tail fairing as separate items. The vertical tail comes in two halves, the port one incorporating the rudder. The wings are split into three parts each (one for the top and two for the bottom), featuring the correct camber; the ailerons and flaps (and even flap actuators!) are separate parts. External stores comprise two R-98s, two R-60s and two UPK-23-250 cannon

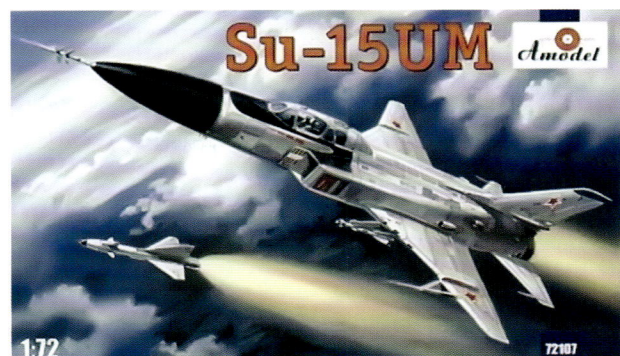

pods. The cockpit is more detailed than the threadbare version in the VES Model kit, but still cries out for an aftermarket replacement. The size A5 six-page instruction booklet again features the aircraft's concise development history and straightforward assembly and painting instructions (with Humbrol enamel numbers as a reference). The high-quality decal sheet features a number of maintenance stencils and insignia/tactical code options for four aircraft, including a Ukrainian Air Force/62nd IAP machine and a camouflaged Soviet Air Force example; however, the painting instructions for the latter are confusing and the veracity of the camouflage scheme is open to doubt. The nose gear unit deserves some criticism, as the wheel axle is too far forward – the levered suspension is not reproduced. Also, the two parts forming the underside of each wing fit into a recess on the upper part (that way an acceptably thin leading edge is provided), so they need to be sanded down a little to make sure they don't stand proud.

Above left: The box top of Amodel's 1:72nd scale Su-15TM.

Above: The box top of Amodel's 1:72nd scale Su-15UM.

Left and below: The Amodel Su-15TM built out of the box by a Russian modeller with the internet alias Mishutka. The model was painted with Tamiya enamels and automotive paint from the rattle can.

Opposite page:

The box top of Trumpeter's 1:72nd scale Su-15UM.

The 1:72nd scale Trumpeter Su-15UM built by a Russian modeller with the internet alias Denis0102.

Two views of an Su-15T converted from Amodel's Su-15TM by Aleksandr Krasnyukov.

AFTERMARKET ITEMS

There are numerous items on the market for those who wish to add realism to their Su-15 model and correct the howlers present in many of the *Flagon* kits. Since the Trumpeters kits are the most widespread, most of the aftermarket parts are designed for these.

In **1:48th scale**, a resin accessory kits manufacturer from Wichita, Kansas, called **Seamless Suckers** offers a set of inlet ducts for the Su-15 *sans suffixe*/Su-15TM/Su-15UM (Ref. No.SS-26). The Ukrainian manufacturer **North Star Models** offers the tail section and engine nozzles/afterburners cast in resin (Ref. No.NS48023). The Russian resin parts company **NeOmega** has a complete Su-15TM cockpit set (Ref. No.C49) and, as a separate item, the correct Sukhoi KS-3/KS-4 ejection seat; they also have a replacement nosewheel well for the Su-15TM/Su-15UM (which is a separate 'tub' on the Trumpeter kits).

Another Russian company, **Equipage**, offers two sets of resin/rubber wheels – a KT-61/3 nosewheel and KT-117 mainwheels for the Su-15 *sans suffixe* and two KN-9 nosewheels and KT-117 mainwheels for the Su-15T/Su-15TM/Su-15UM, while **Scale Aircraft Conversions** from Dallas, Texas, offers a set of landing gear struts cast in white metal (Ref. No.48109). Two replacement resin radomes for the Su-15TM and Su-15UM with the correct shape and correctly positioned joint line are available from **QuickBoost** (Ref. No.QB 48 207) and **Loon Models** (Ref. No.LO48201); they have a lot of resin inside, providing extra ballast in the nose. However, the Loon Models nose has the same flaw as the stock Trumpeter radome – the nose ends up being way too long! The QuickBoost radome creates a different problem; the forward fuselage is too narrow for it, and if you install it, the result is an ugly gap between the fuselage halves which requires copious filling and sanding!

QuickBoost also has a set of cooling air scoops for the Su-15 *sans suffixe* (Ref. No.QB 48 203) and a resin/photo-etched air data boom for the Su-15TM/Su-15UM (Ref. No.QB 48 214) which comes complete with an assembly jig to assist the complex assembly procedure, making sure that the pitch/yaw vanes and aerials carried on the boom are correctly glued at 90° to each other. The Czech company **Eduard Models** offers PE parts sets for the cockpit, main gear doors and airbrakes of the Su-15 *sans suffixe* (Ref. No.48400) and the Su-15TM (Ref. No.48402), as well as the engine afterburners (Ref. No.48404); another PE parts set for the Su-15TM and Su-15UM comprising intake and afterburner parts, wing fences, pitots etc. is available from **Part** (Ref. No.S48-104).

In **1:72nd scale**, the offer is broadly similar. Replacement radomes are available for the Su-15 *sans suffixe* from the Czech resin parts company **Pavla Models** (Ref. No.U72-95) and for the Su-15TM/Su-15UM from Pavla (Ref. No.U72-93) and **QuickBoost** (Ref. No.QB 72 173). however, the same problem arises; the forward fuselages of the Trumpeter and Pioneer 2/PM Model kits are too narrow, and if you insert the replacement resin radome the fuselage halves won't go together properly! Additionally, QuickBoost has a resin/PE air data boom for the Su-15TM/Su-15UM (Ref. No.QB 72 176).

Pavla also offers correct vertical tails for the Su-15 *sans suffixe* (Ref. No.U72-94) and the Su-15TM/Su-15UM (Ref. No.U72-92), a Su-15 *sans suffixe*/Su-15TM cockpit set including two correctly shaped vacuform canopies with or without rear view mirror (Ref. No.C72075), a similar set for the Su-15TM (Ref. No.C72073), a resin-only cockpit set for the Su-15UM (Ref. No.C72081), air intakes for the Su-15UM (Ref. No.U72-105) and a set of cooling air scoops for the Su-15TM (Ref. No.U72-106). **NeOmega** again has a KS-3/KS-4 ejection seat – which, incidentally, does not fit into the cockpit of the Amodel kit. NeOmega also offers a complete Su-15TM-to-Su-15 *sans suffixe* conversion kit for the VES Model kit comprising wings (with integrally moulded wing fences and wheel well interiors), vertical tail, radome and a rather basic vacuform canopy – but, unfortunately, no replacement single-wheel nose gear unit, which has to be sourced elsewhere.

This page:
The cockpit parts and airbrakes from Eduard's 1:48 PE set for the Su-15TM

Opposite, clockwise from top left:
Northstar Models' 1:48th scale resin nozzles.

NeOmega 1:72nd scale Su-15TM to Su-15 conversion kit

Pavla's 1:72nd scale Su-15 *sans suffixe* resin cockpit and vacuform canopy set.

Pavla's 1:72nd scale Su-15 *sans suffixe* vertical tail.

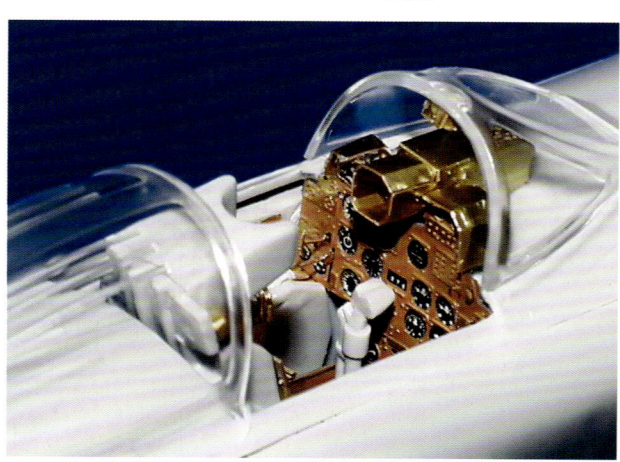

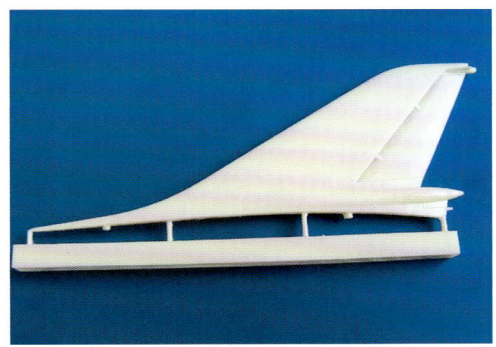
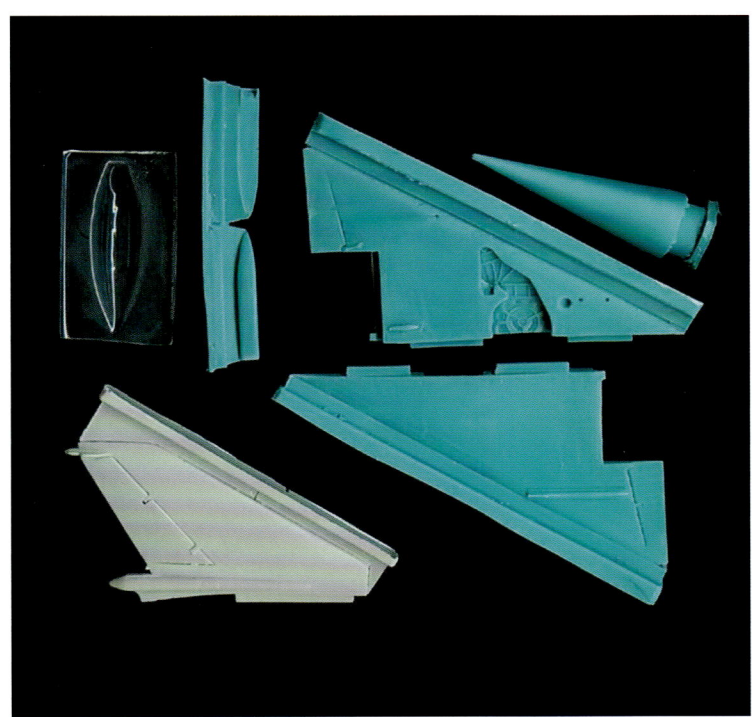

Equipage has resin/rubber wheels for the Su-15 *sans suffixe* (Ref. No.72187) and also the Su-15T/Su-15TM/Su-15UM (Ref. No.72188), while the Czech company **Aires Hobby Models** offers all-resin wheels and painting masks for the same aircraft types (Ref. No.7218). PE parts include a set of cockpit parts from **Eduard Models** (Ref. No.SS311) and a set of gear doors and assorted panels and aerials from the same company (Ref. No.73311).

Finally, the Russian company **Begemot Decals** ('Hippopotamus') offers high-quality decal sheets for the Su-15 in both scales.

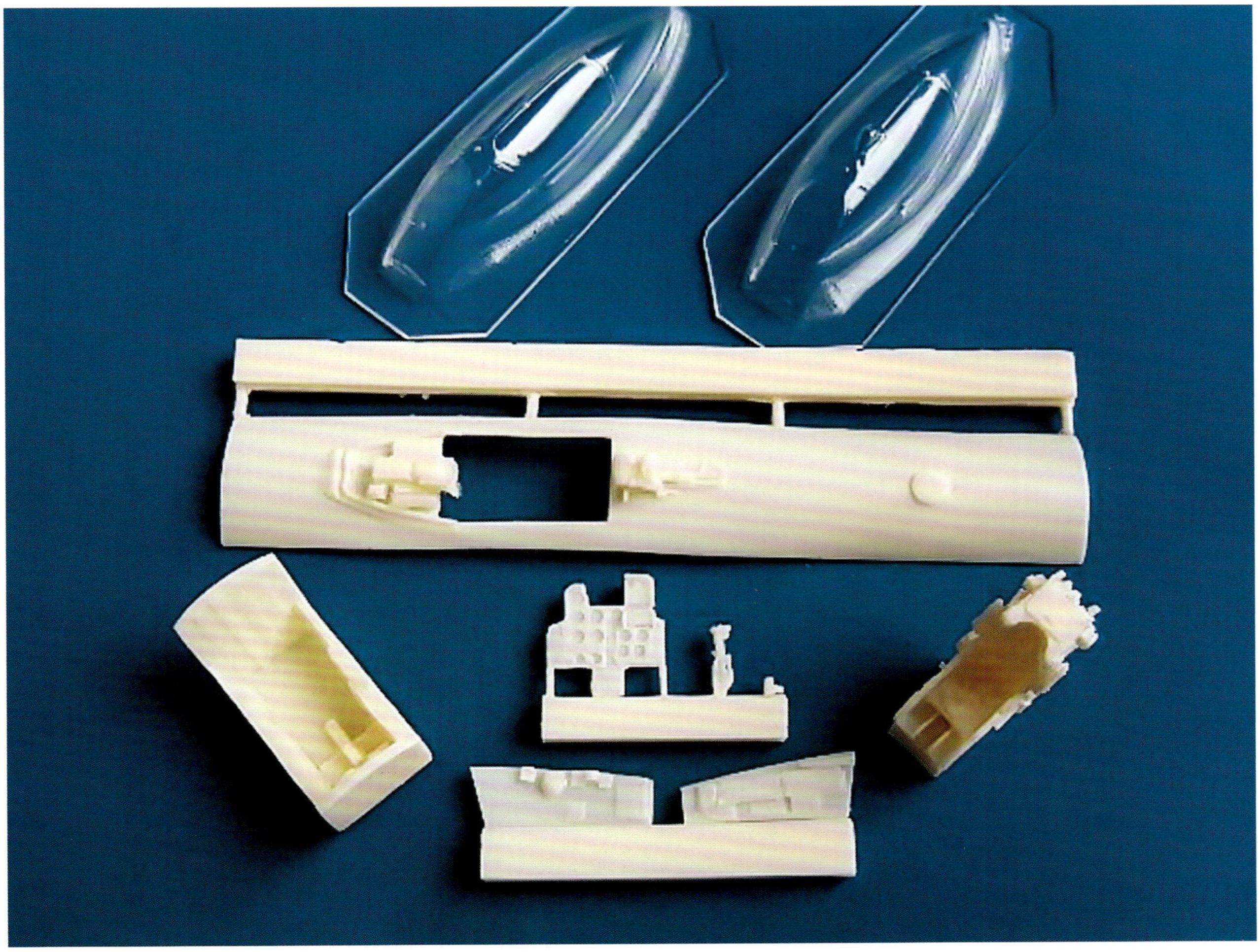

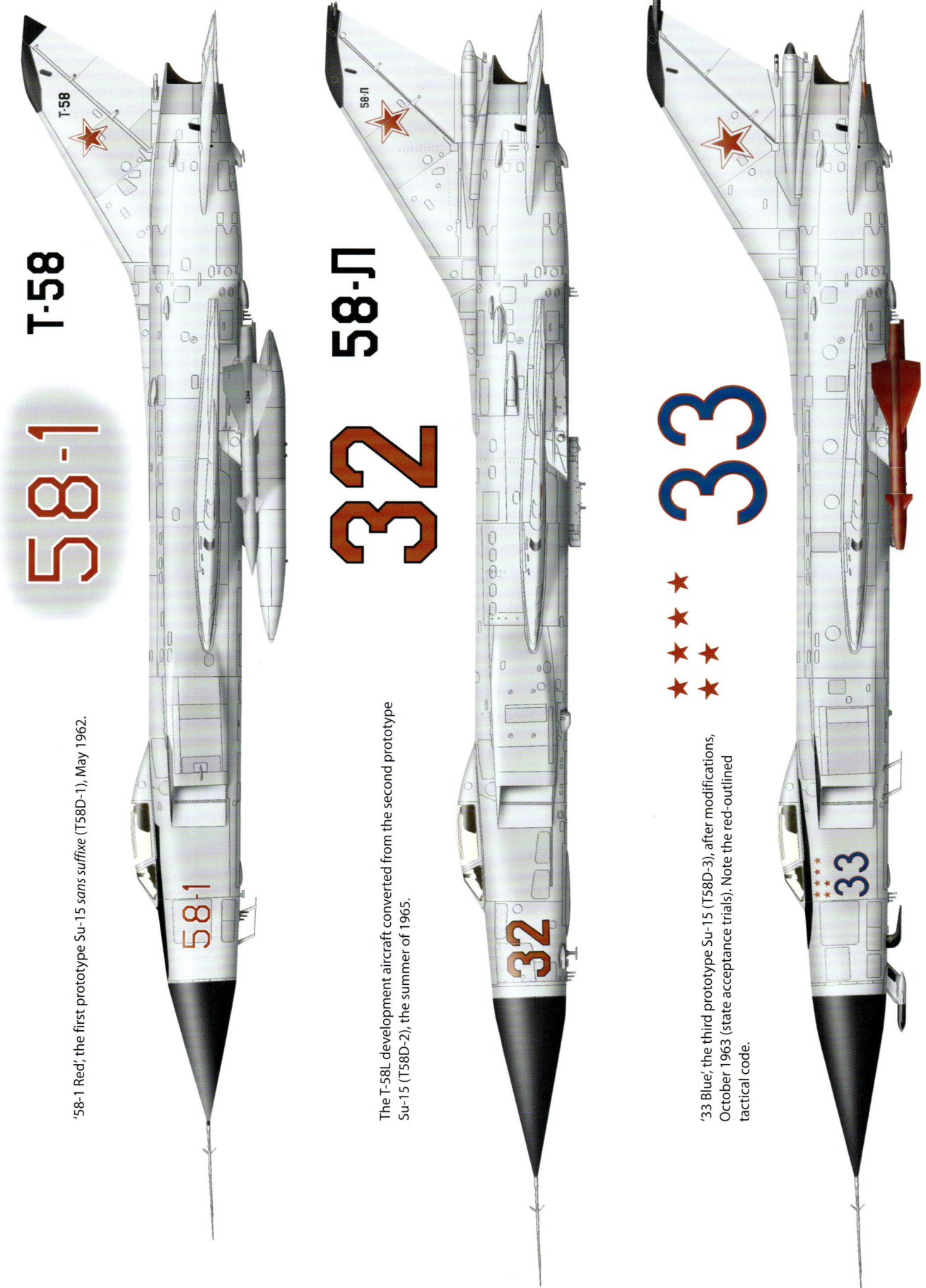

'58-1 Red', the first prototype Su-15 sans suffixe (T58D-1), May 1962.

The T-58L development aircraft converted from the second prototype Su-15 (T58D-2), the summer of 1965.

'33 Blue', the third prototype Su-15 (T58D-3), after modifications, October 1963 (state acceptance trials). Note the red-outlined tactical code.

'34 Red' (c/n 0015301), the first pre-production Su-15 sans suffixe, February 1966.

Black-painted Su-15 '47 Red' flown by Vladimir S. Il'yushin at the Moscow-Domodedovo airshow on 9th July 1967.

The T-58VD as it appeared at the Moscow-Domodedovo airshow.

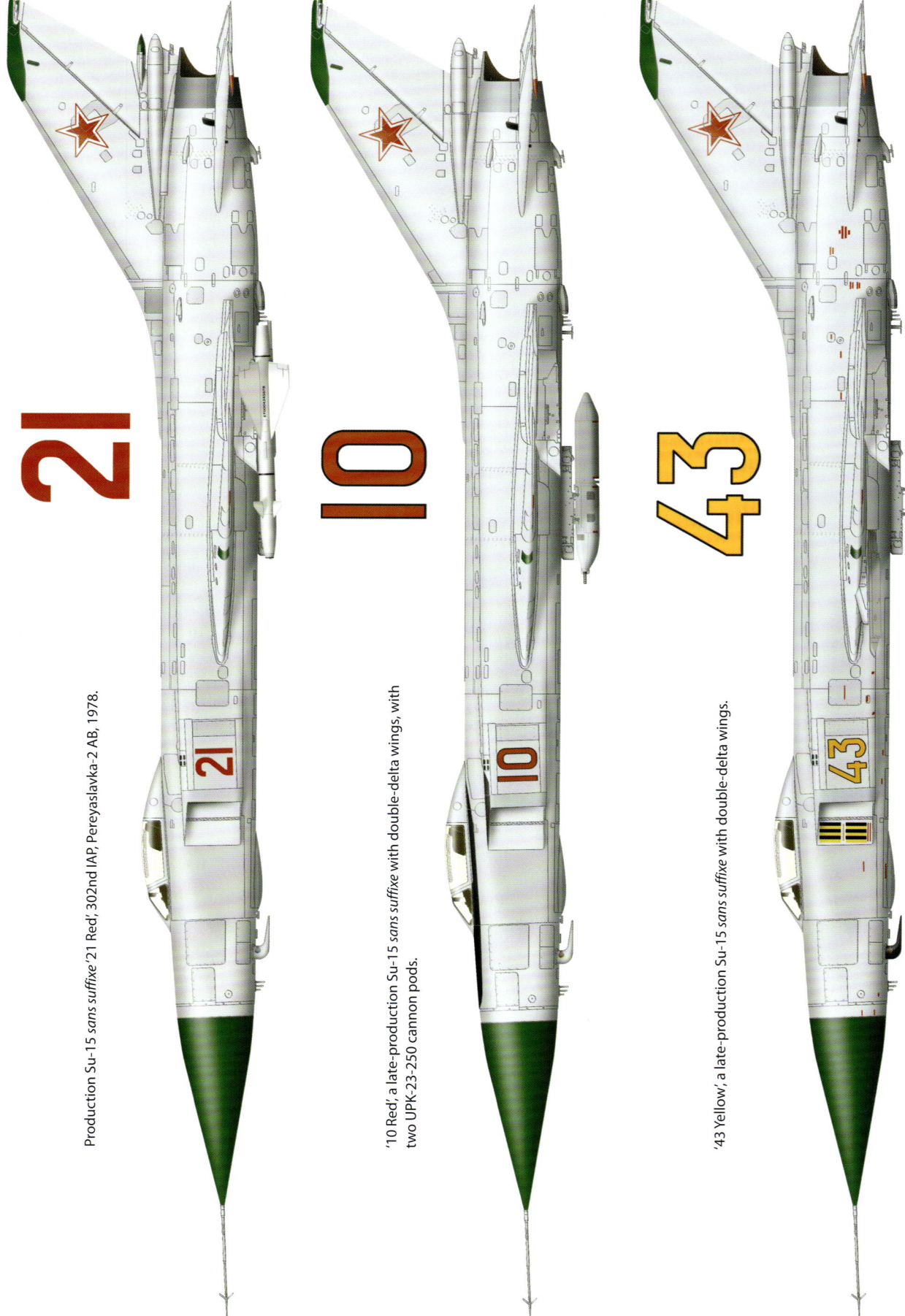

Production Su-15 *sans suffixe* '21 Red', 302nd IAP, Pereyaslavka-2 AB, 1978.

'10 Red', a late-production Su-15 *sans suffixe* with double-delta wings, with two UPK-23-250 cannon pods.

'43 Yellow', a late-production Su-15 *sans suffixe* with double-delta wings.

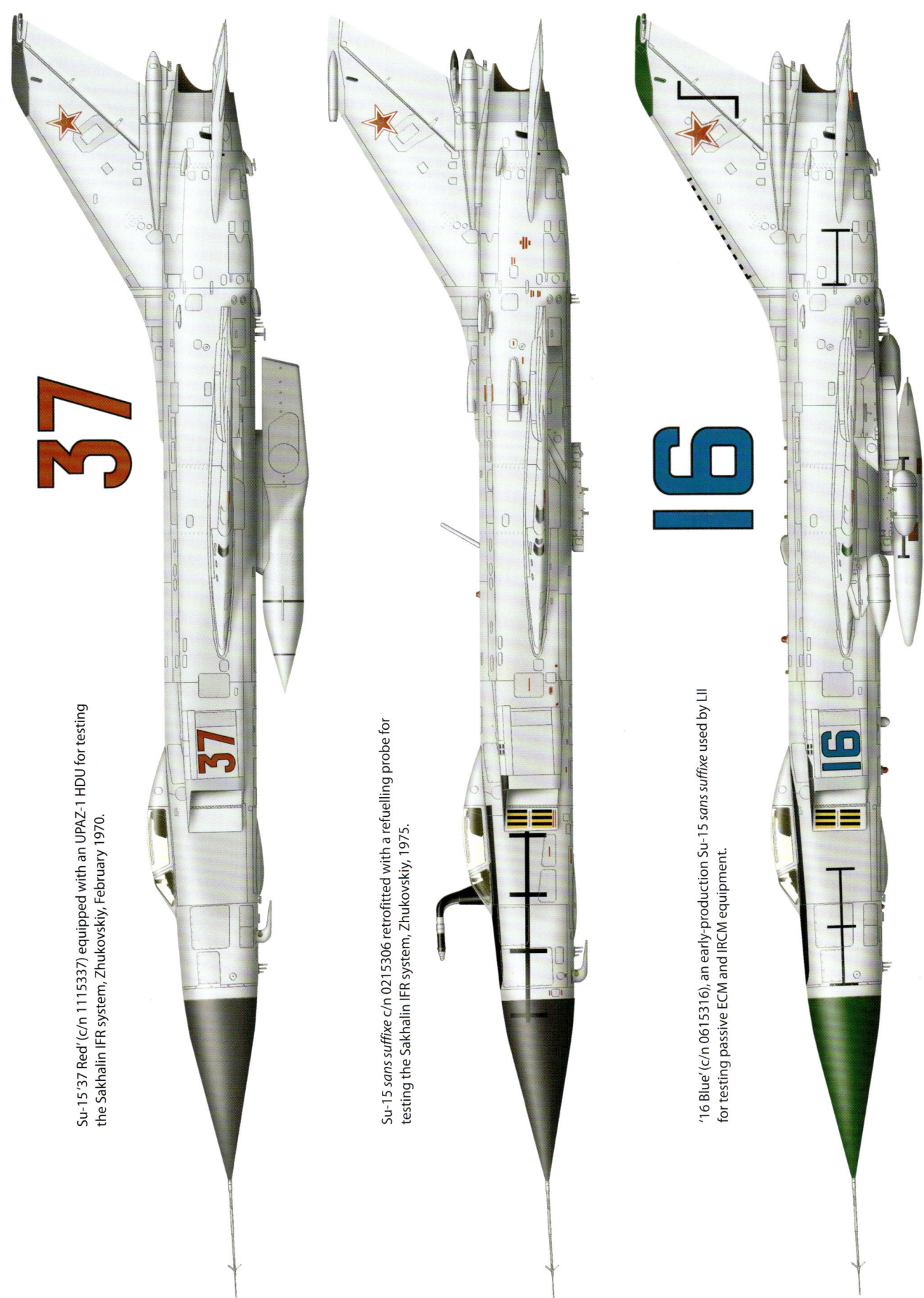

Su-15 '37 Red' (c/n 1115337) equipped with an UPAZ-1 HDU for testing the Sakhalin IFR system, Zhukovskiy, February 1970.

Su-15 sans suffixe c/n 0215306 retrofitted with a refuelling probe for testing the Sakhalin IFR system, Zhukovskiy, 1975.

'16 Blue' (c/n 0615316), an early-production Su-15 sans suffixe used by LII for testing passive ECM and IRCM equipment.

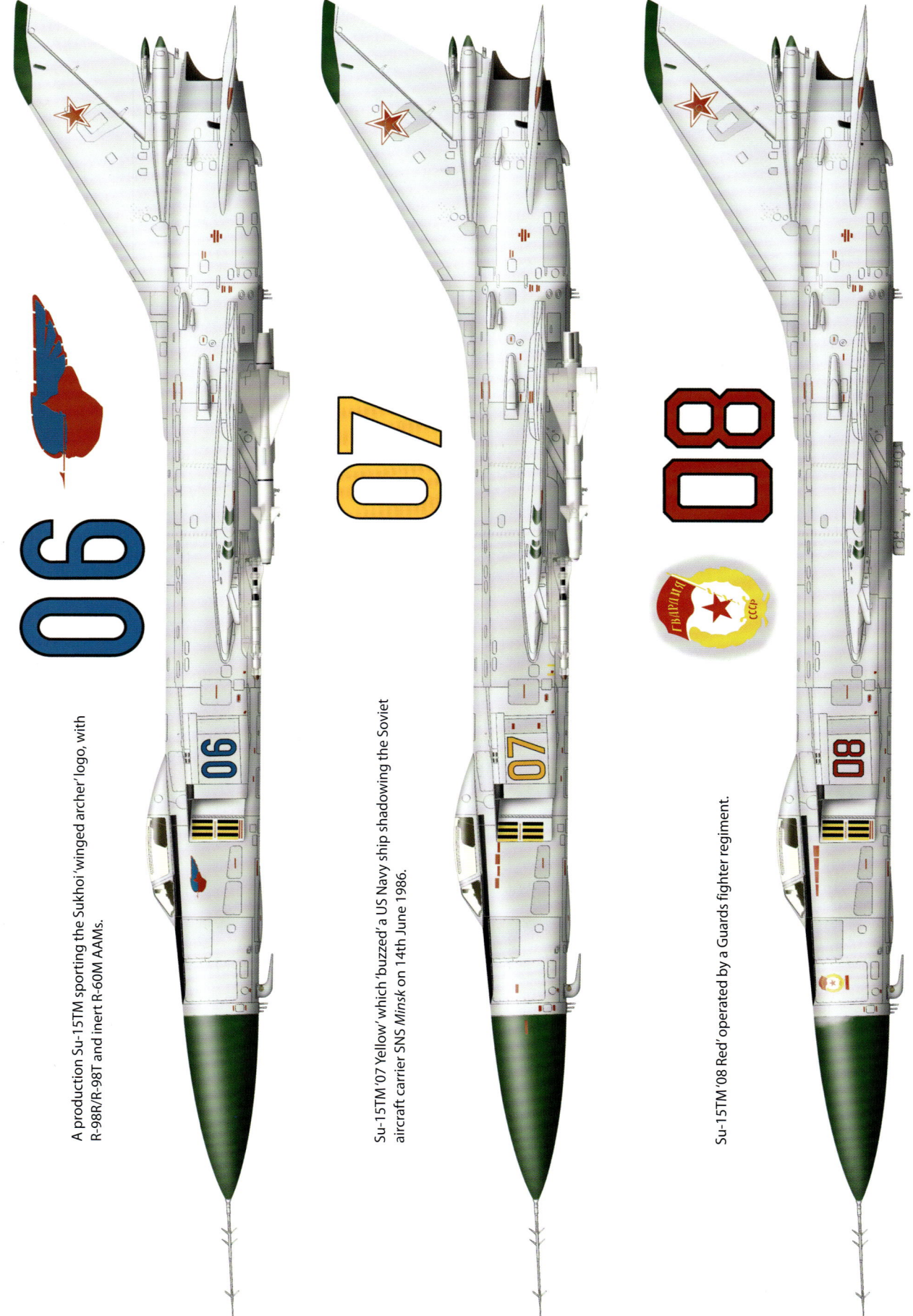

A production Su-15TM sporting the Sukhoi 'winged archer' logo, with R-98R/R-98T and inert R-60M AAMs.

Su-15TM '07 Yellow' which 'buzzed' a US Navy ship shadowing the Soviet aircraft carrier SNS *Minsk* on 14th June 1986.

Su-15TM '08 Red' operated by a Guards fighter regiment.

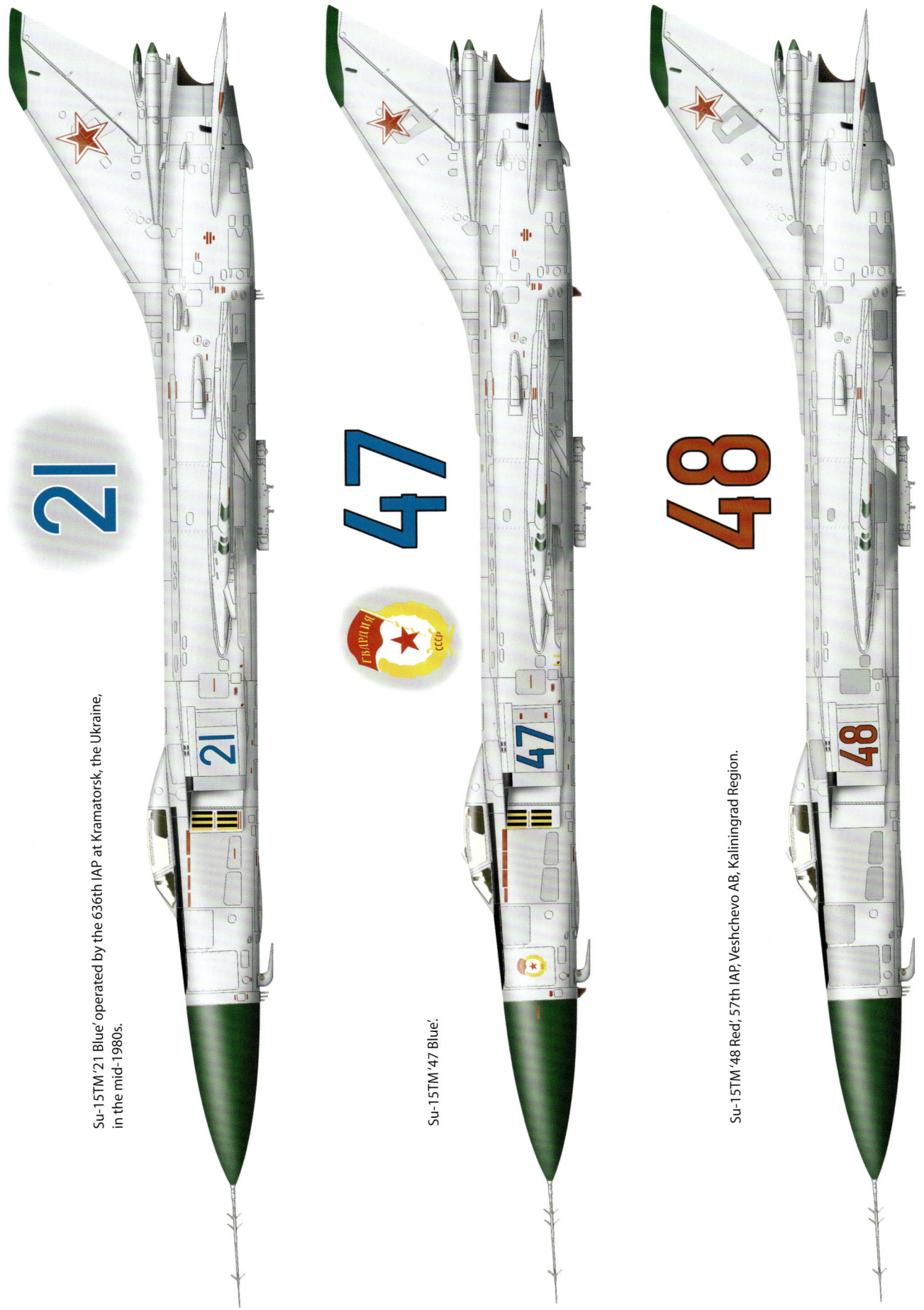

Su-15TM '21 Blue' operated by the 636th IAP at Kramatorsk, the Ukraine, in the mid-1980s.

Su-15TM '47 Blue'.

Su-15TM '48 Red', 57th IAP, Veshchevo AB, Kaliningrad Region.

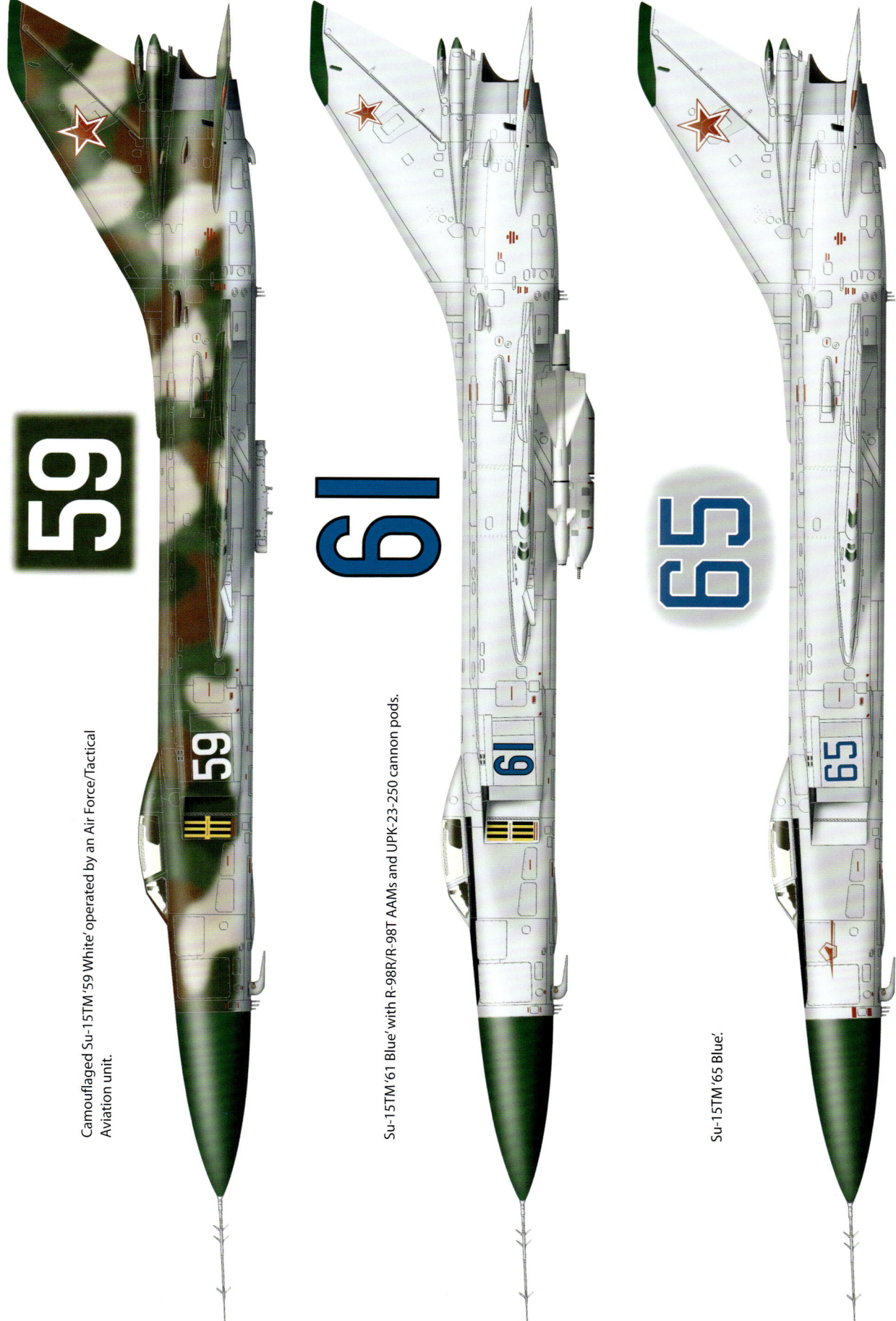

Camouflaged Su-15TM '59 White' operated by an Air Force/Tactical Aviation unit.

Su-15TM '61 Blue' with R-98R/R-98T AAMs and UPK-23-250 cannon pods.

Su-15TM '65 Blue'.

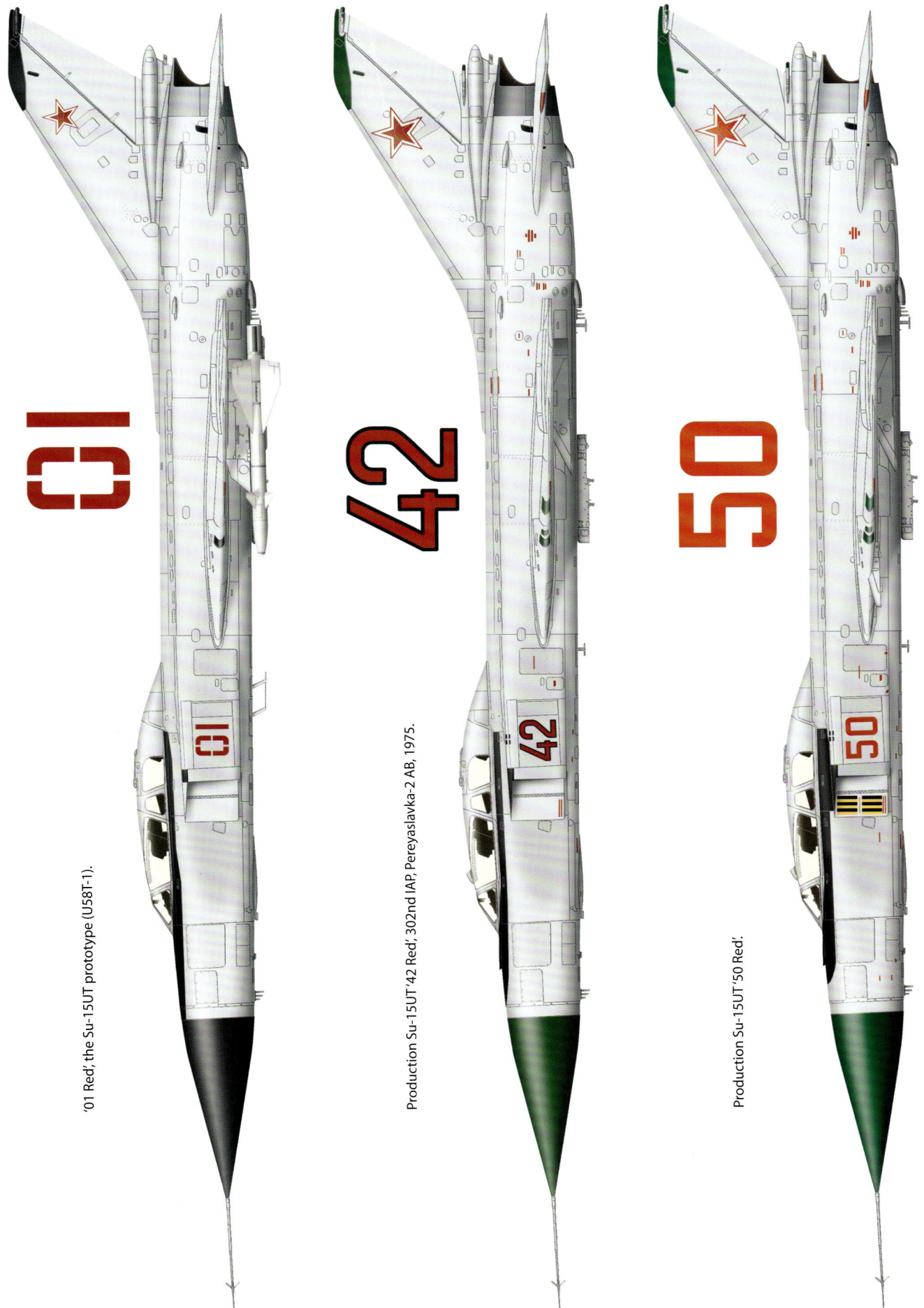

'01 Red', the Su-15UT prototype (U58T-1).

Production Su-15UT '42 Red', 302nd IAP, Pereyaslavka-2 AB, 1975.

Production Su-15UT '50 Red'.

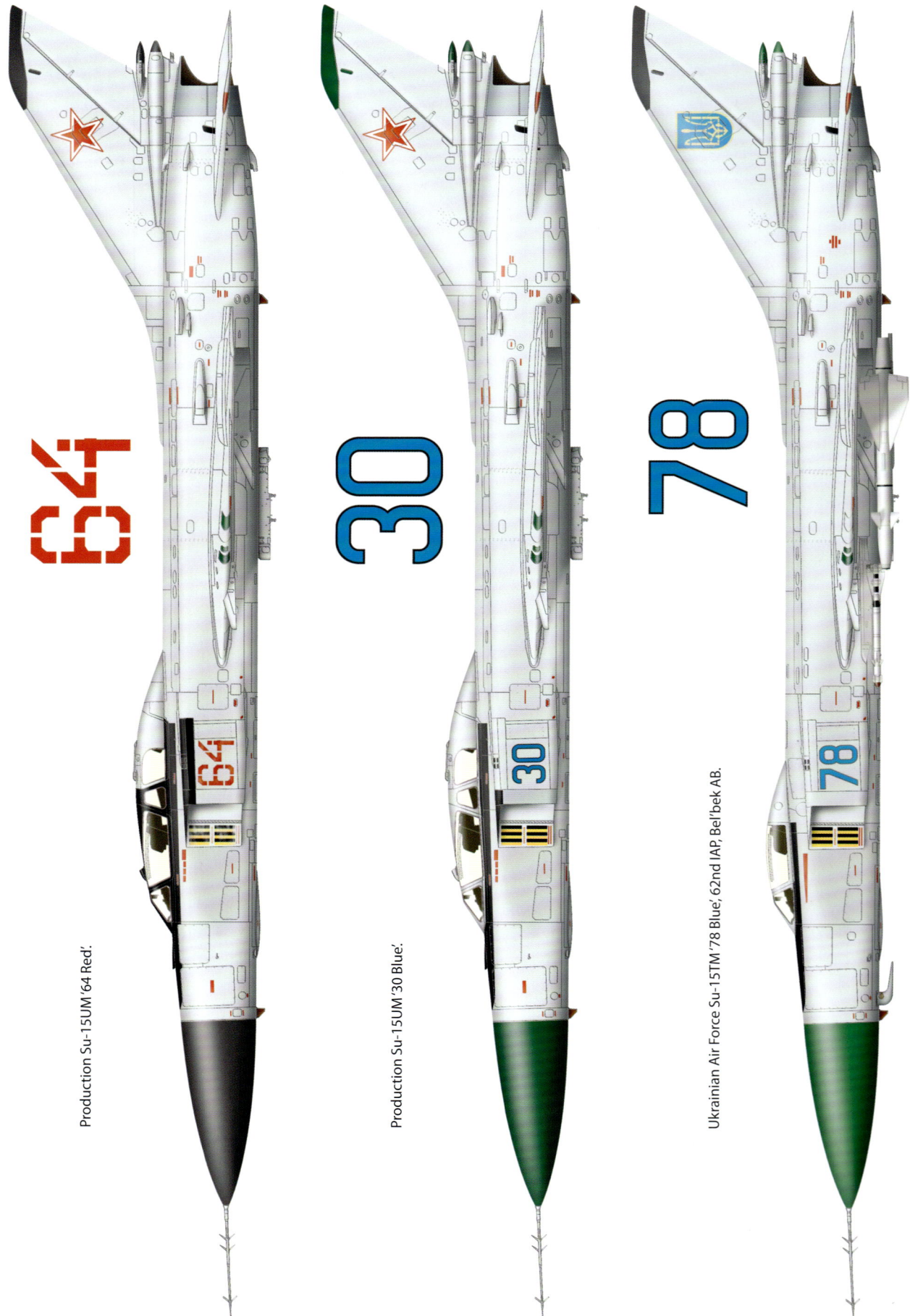

Production Su-15UM '64 Red'.

Production Su-15UM '30 Blue'.

Ukrainian Air Force Su-15TM '78 Blue', 62nd IAP, Bel'bek AB.

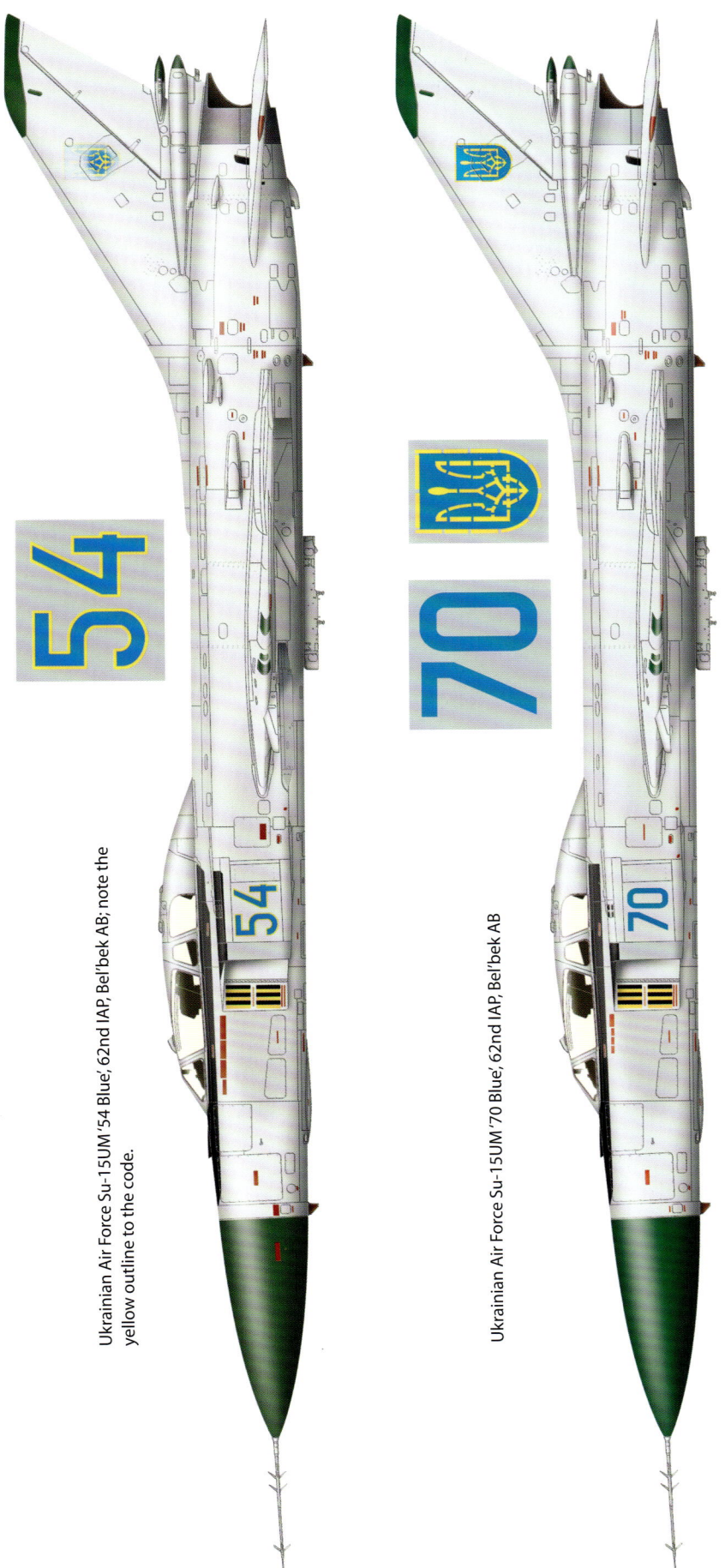

Ukrainian Air Force Su-15UM '54 Blue', 62nd IAP, Bel'bek AB; note the yellow outline to the code.

Ukrainian Air Force Su-15UM '70 Blue', 62nd IAP, Bel'bek AB

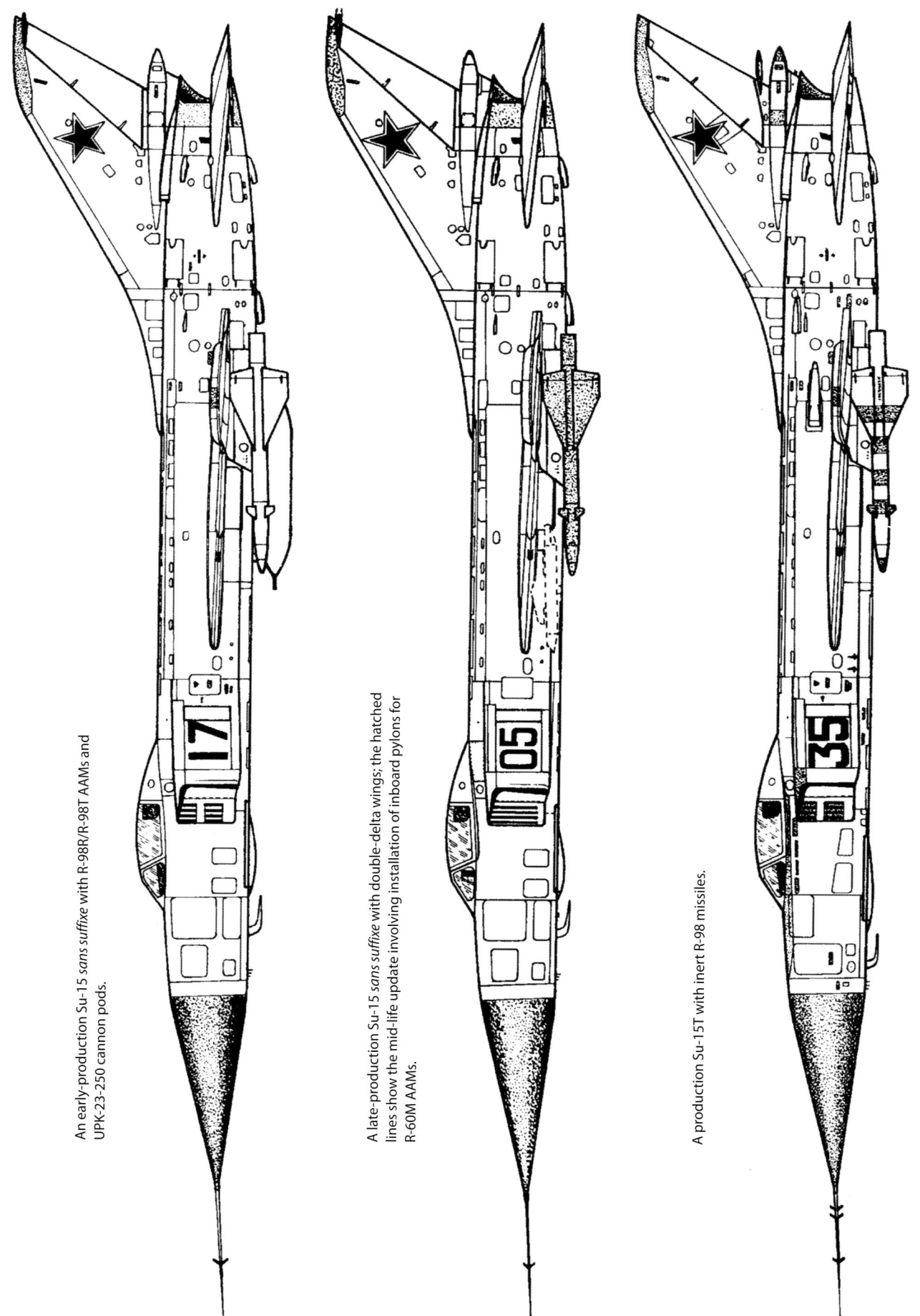

An early-production Su-15 sans suffixe with R-98R/R-98T AAMs and UPK-23-250 cannon pods.

A late-production Su-15 sans suffixe with double-delta wings; the hatched lines show the mid-life update involving installation of inboard pylons for R-60M AAMs.

A production Su-15T with inert R-98 missiles.

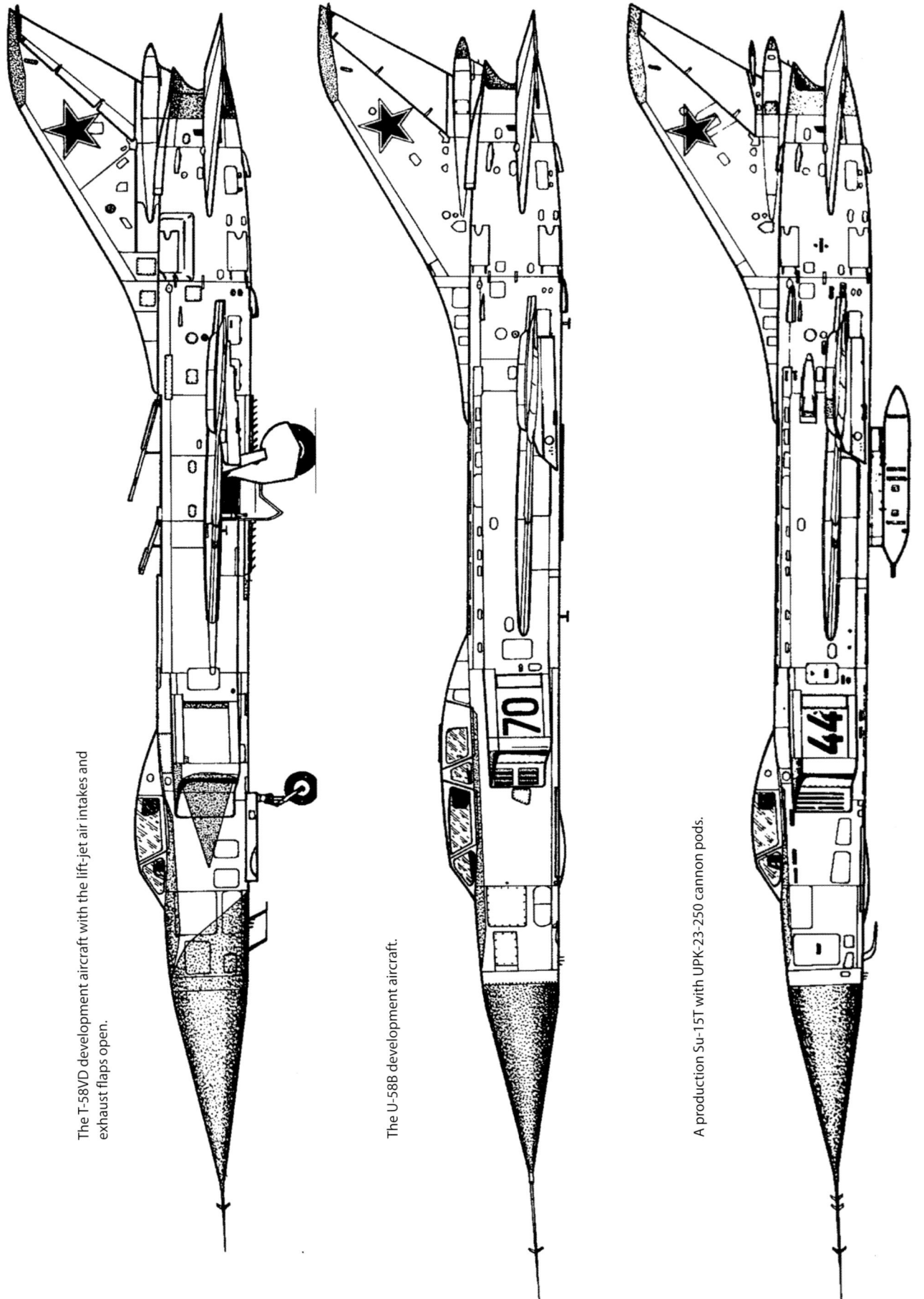

The T-58VD development aircraft with the lift-jet air intakes and exhaust flaps open.

The U-58B development aircraft.

A production Su-15T with UPK-23-250 cannon pods.

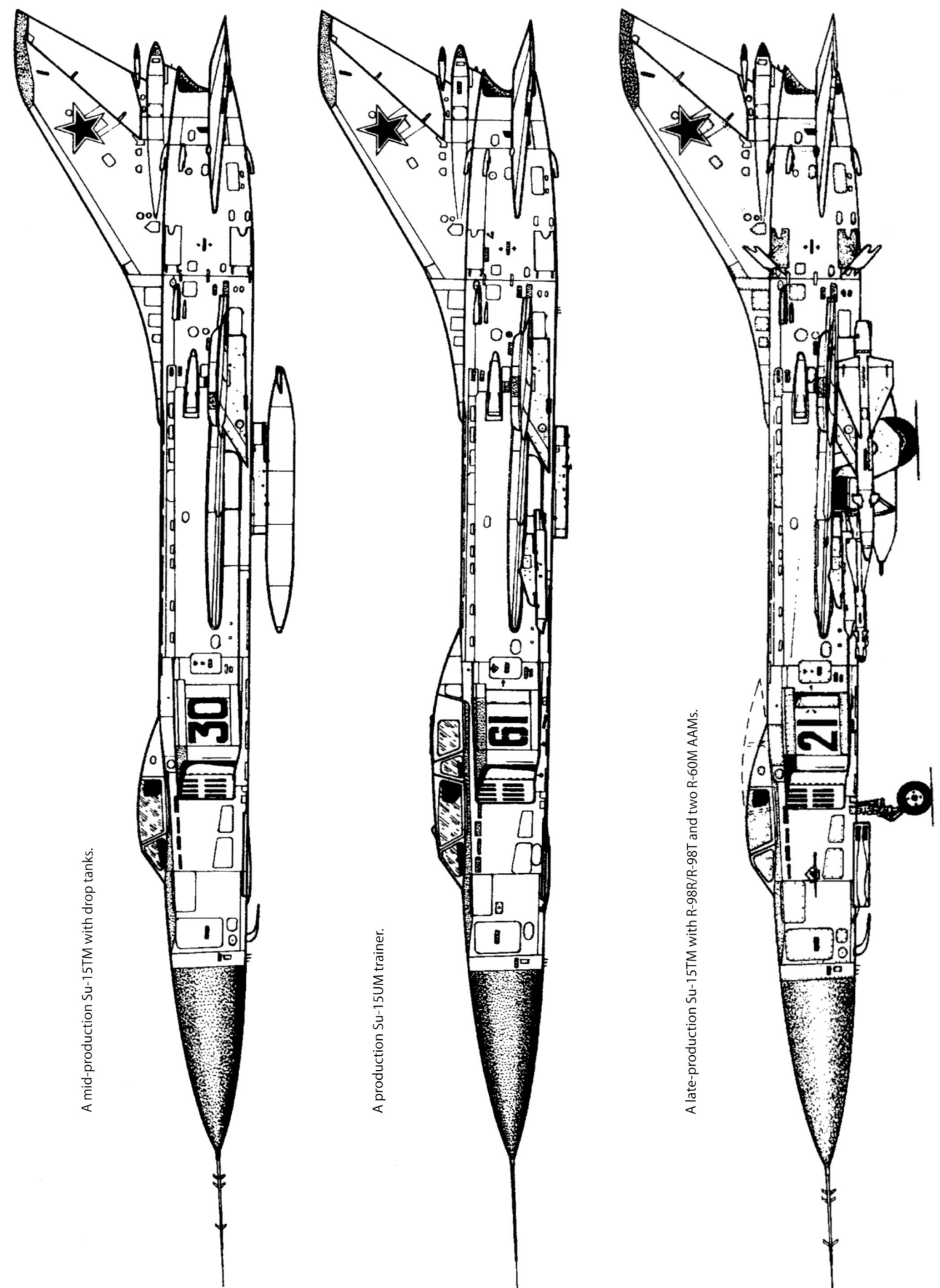

A mid-production Su-15TM with drop tanks.

A production Su-15UM trainer.

A late-production Su-15TM with R-98R/R-98T and two R-60M AAMs.

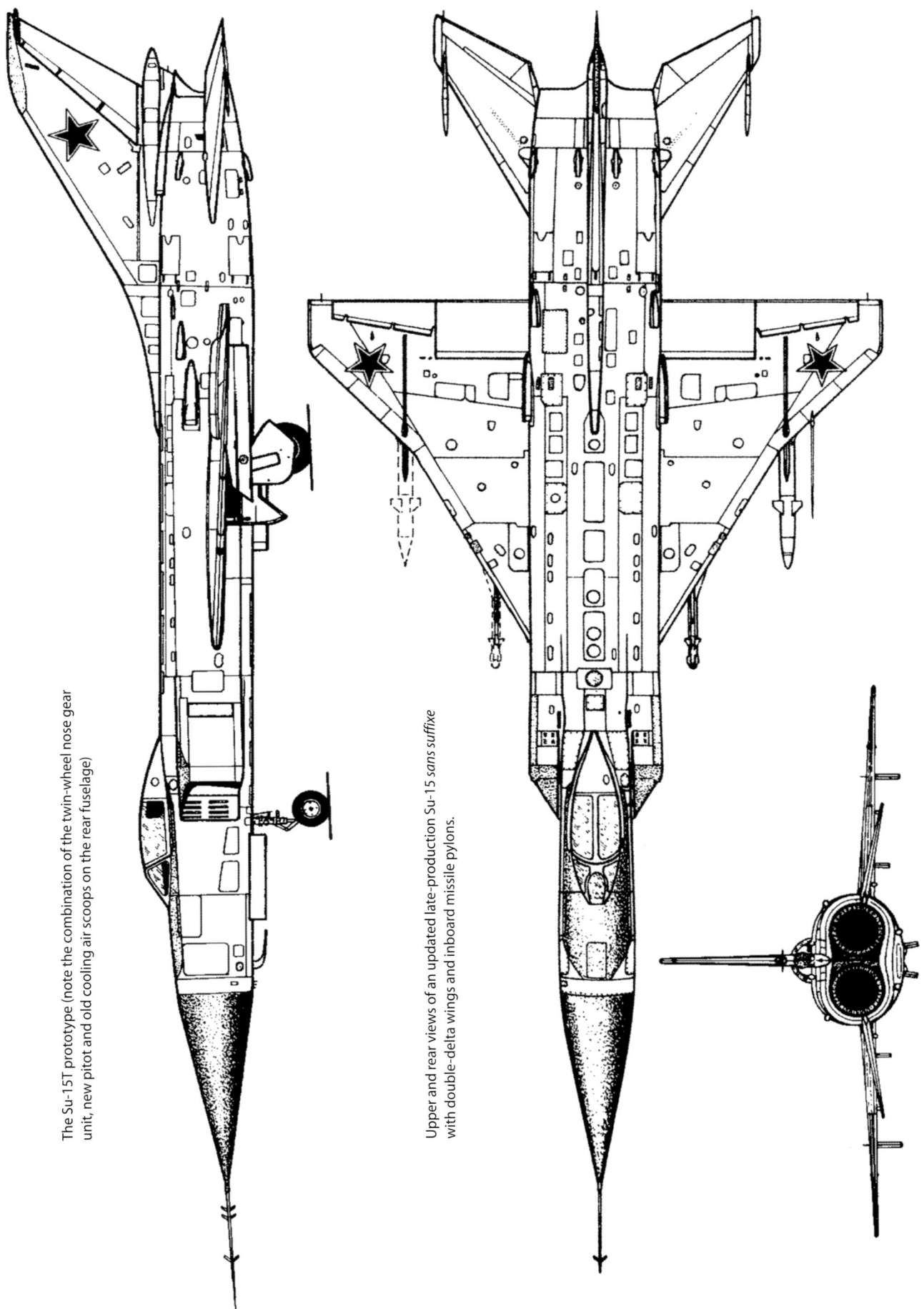

The Su-15T prototype (note the combination of the twin-wheel nose gear unit, new pitot and old cooling air scoops on the rear fuselage)

Upper and rear views of an updated late-production Su-15 *sans suffixe* with double-delta wings and inboard missile pylons.

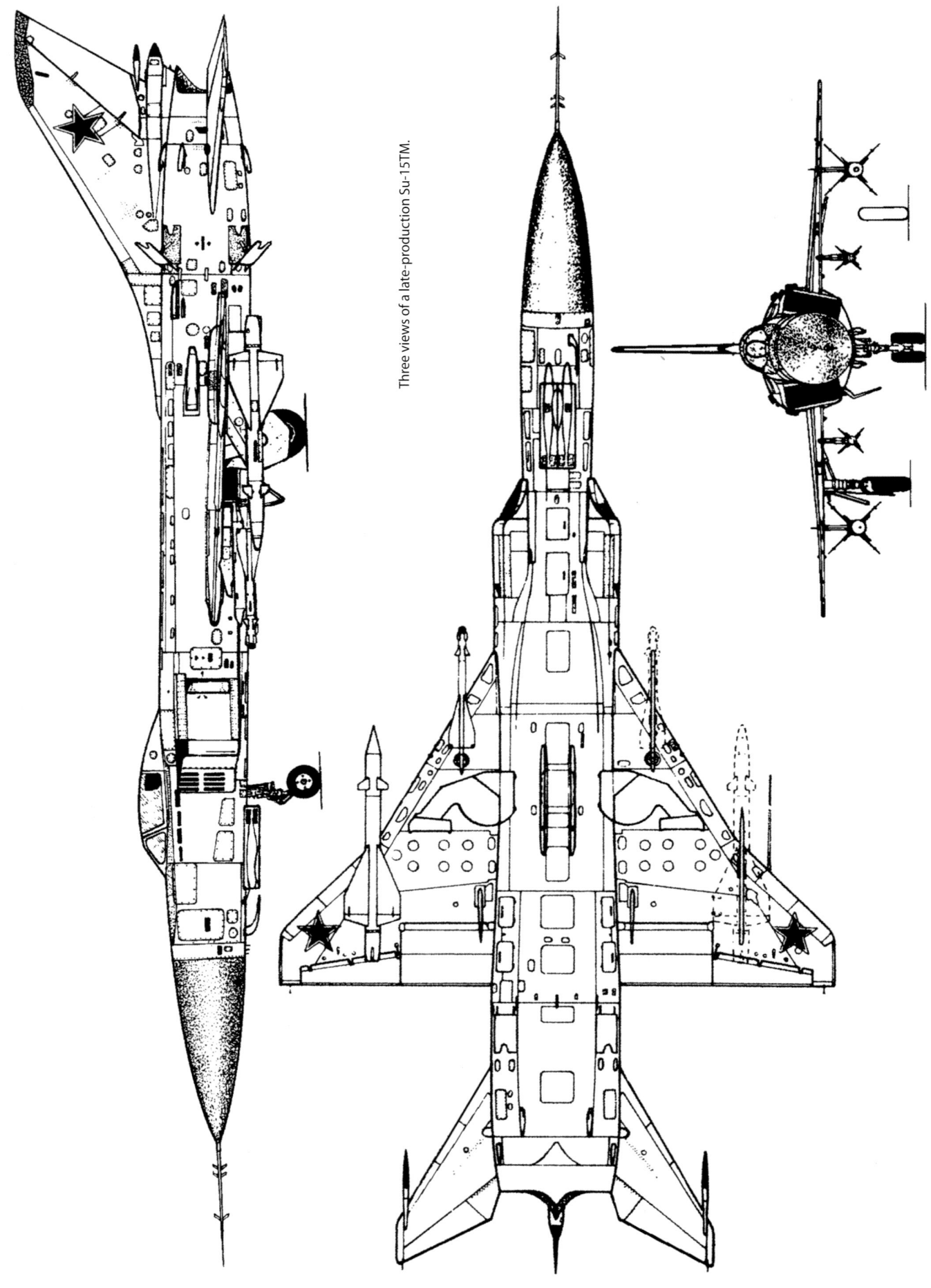

Three views of a late-production Su-15TM.